Coal: 1985 and Beyond

AF719254

Coal: 1985 and Beyond

A perspective study of the coal industry in Europe and North America prepared for the Coal Committee of the United Nations Economic Commission for Europe.

Published by

PERGAMON PRESS

for the

UNITED NATIONS

U.K.	Pergamon Press Ltd., Headington Hill Hall, Oxford OX3 0BW, England
U.S.A.	Pergamon Press Inc., Maxwell House, Fairview Park, Elmsford, New York 10523, U.S.A.
CANADA	Pergamon of Canada Ltd., 75 The East Mall, Toronto, Ontario, Canada
AUSTRALIA	Pergamon Press (Aust.) Pty. Ltd., 19a Boundary Street, Rushcutters Bay, N.S.W. 2011, Australia
FRANCE	Pergamon Press SARL, 24 rue des Ecoles, 75240 Paris, Cedex 05, France
FEDERAL REPUBLIC OF GERMANY	Pergamon Press GmbH, 6242 Kronberg-Taunus, Pferdstrasse 1, Federal Republic of Germany

---

Copyright © 1978 United Nations

All Rights Reserved. No part of this publication may be reproduced, stored in a retrieval system or transmitted in any form or by any means: electronic, electrostatic, magnetic tape, mechanical, photocopying, recording or otherwise, without permission in writing from the copyright holders.

First edition 1978

HD 9555 .A5 C54 1978

British Library Cataloguing in Publication Data

Economic Commission for Europe Coal Committee
Coal, 1985 and beyond
1 Coal mines and mining — Europe
I Title
338 2'7'2094 HD9555 77-30437

ISBN 0-08-022409-1

In order to make this volume available as economically and as rapidly as possible the author's typescript has been reproduced in its original form. This method unfortunately has its typographical limitations but it is hoped that they in no way distract the reader.

Printed in Great Britain by Page Bros. (Norwich) Ltd., Norwich and London

CONTENTS

PREFACE

Origin of the Study

The Economic Commission for Europe (ECE) is a United Nations regional instrument used by its member governments (European States, United States of America and Canada) for

- the promotion of international trade
- scientific and technological co-operation
- policy making aimed at long-term economic growth
- improvement of the environment.

The present publication which is the first of a series, is released to the public domain with a view to promoting the dissemination of information on a subject of crucial importance - the perspectives of the coal industry in the ECE region. This topic had been examined between December 1974 and July 1977 by the Working Party on Coal Trade, a subsidiary organ of the Coal Committee of ECE. The material submitted to, and considered by the Working Party was given general distribution, as customary under the responsibility of the secretariat, by the Coal Committee at its 73rd session held in September 1977.

Purpose Pursued

The purpose of the study is to identify the consequences of the energy crisis, or of any other new developments since 1973/74, on the demand for and supply of coal in the ECE region (Europe and North America), and to describe related government policies as formulated during 1975-1976. While it is obvious that these policies are subject to further change, it is not expected that any such changes would significantly affect the conclusions of the present study with regard to the medium-term.

Sources Used

The study aims at describing government policies. Accordingly it is essentially based on government replies in response to the requests of the Working Party on Coal Trade or of any other subsidiary body of ECE. Other official documentation and information published in the specialized press has been used as well.

Definitions, Conversion Factors

Figures given by the national authorities may not necessarily be internationally comparable. Whereas national figures are reproduced unchanged in Part I, an effort was made to render them internationally comparable in Parts II - V according to the definitions, conversion factors and time series reproduced in the UN Statistical Series "World Energy Supplies". The main remaining discrepancies lie in the inclusion or exclusion of stock changes and of recovered slurry.

Symbols Used

• = secretariat estimates

... = no information available
– = nil, or less than half of the unit employed
t = metric ton
tce = ton of coal equivalent

Concluding Date

The report contains Government statements and secretariat analyses as of 1st August 1977.

PART I

PERSPECTIVES OF THE COAL INDUSTRY IN THE VARIOUS COUNTRIES[1]

A. AUSTRIA

The shock of the threat to the supply of petroleum in the autumn of 1973 and the acute soaring of prices for imported crude oil undoubtedly necessitated the reflection of change of views on the energy sector, which could result in structural changes in the near and far future. Currently, the effects on the future development of lignite mining in Austria on a short-term basis and with respect to other development components (savings of liquid fuels, trade recession, atmospheric conditions, etc.) are being appraised for individual symptomatic indications. A long-term forecast drawn up on a scientific basis could far better illustrate future relations following the crisis. Difficulties have arisen in regard to an energy programming after the energy crisis; an exact forecast of the extent and type of effect of the crisis on coal-mining in Austria will not be available for some time.

(a) Definition of the Significance of Solid Fuels in the Overall Concept of Energy

The statistics outlined below provide an overall view of the relative importance of solid fuels and the development of their percentage share in the production and distribution of total energy from 1969-1974 in Austria.

TABLE I.1 The Relative Importance of Coal

Energy production/consumption	1969	1970	1971	1972	1973	1974
Total production	26.0	24.1	20.6	20.2	18.6	19.6
Availability	18.0	19.1	13.0	14.5	13.4	13.9
Final consumption	20.6	19.9	15.8	14.3	13.6	14.1
Transport sector	18.1	16.9	13.3	6.8	5.2	4.1
Industry	23.0	20.2	18.3	15.2	14.3	16.4
Domestic fuel	18.7	20.6	13.9	15.9	15.7	14.9

During this period the total energy production rose by 25.1 per cent while the import dependency increased from 53.2 per cent in 1969 to 61.7 per cent in 1974.

1/ As transmitted by the respective Governments

In 1975 imports were 2.6 mill.t of bituminous coal, 0.68 mill.t of lignite and lignite briquettes and 0.9 mill.t of bituminous cokes. The domestic coke production amounted to 1.6 mill.t and lignite production to 3.4 mill.t. Domestic lignite was principally consumed by power stations and long-distance heating stations (2.6 mill.t) and to a much lesser extent by industry (0.3 mill.t) and other consumers (domestic) (0.6 mill.t).

The short-term changes in 1973 and 1974 indicate a slight increase in the share of solid fuels (4 per cent) against a decrease in the share of petroleum products (-11 per cent) but this is almost exclusively due to the increased importation of coking coal. The simultaneous increase in natural gas requirements (+9 per cent) and hydro power (+12 per cent) lead to the assumption that demand will primarily revert from petroleum to natural gas and electricity and will only result in a minimum increase in the demand for coal.

With regard to indigenous lignite, no signs of an improved sales potential became evident as a result of the crisis. The demand for and the production of domestic lignite are considered to be receding. However, domestic coal mining is shown to be of considerable importance also in the future, particularly because domestic coal is available, import risks are nil, the balance of payments is not encumbered and because domestic coal represents a price-controlling and stabilizing factor.

(b) Forecast of Demand According to Consumer Groups, Qualities and New Sale Markets

The "Energieplan 1976" contains the following forecasts:

TABLE I.2 Development and Structure of Gross Inland Energy Consumption

	1974 (actual)	1980	1985	1990
	in 10^9 kcal			
Total	208.761	260.785	319.354	386.778
of which hard coal	28.525	27.734	29.323	30.226
brown coal	14.693	13.391	11.438	10.451
liquid fuels	106.792	138.852	153.895	186.900
natural gas	36.260	41.510	61.320	71.820
hydropower	21.931	27.615	32.977	37.149
nuclear power	0	10.913	29.281	48.727
others	560	770	1.120	1.505
	in per cent			
Total	100.0	100.0	100.0	100.0
of which hard coal	13.7	10.6	9.2	7.8
brown coal	7.0	5.1	3.6	2.7
liquid fuels	51.1	53.3	48.2	48.3
natural gas	17.4	15.9	19.2	18.6
hydropower	10.5	10.6	10.3	9.6

TABLE I.2 (cont'd)

nuclear power	0	4.2	9.2	12.6
others	0.3	0.3	0.3	0.4

As regards new application potential and market areas an expansion of coke production from imported bituminous coal in coking plants is not being considered at present. There is no plan to extend the coal drying plant which partly treats lignite won at home.

However, the development of new fields and the guarantee of stable markets for an additional 25 mill.t of workable coals permitted the construction of a new thermal power station with a capacity of approximately 300 MW. The initial planning of this is already underway. Similar efforts are also being made with regard to other coal mines and, depending on favourable results, the erection of a further thermal power station (150 MW).

In addition, a large concern supported by the Research Promotion Fund is currently testing the deformability of lignite at low-temperature carbonization under increased pressure.

The potential of underground gasification of lignite with the application of newly developed methods is still in the testing stage.

TABLE I.3 Coal Consumption by Sectors

in 10^9 kcal

	1974 actual	1980	1985	1990
1. Hard coal				
Industry	546	245	175	105
Transport	763	175	98	91
Small consumers	2.121	2.100	1.750	1.750
Conversion into				
- coke	16.268	16.100	16.100	16.100
- electricity	560	70	70	70
Total	20.258	18.690	18.193	18.116
2. Brown coal				
Industry	1.295	1.085	630	350
Transport	70	35	28	21
Small consumers	4.312	3.500	2.800	2.100
Energy sector	161	105	70	70
Transformation into				
- manufactured gas	182	0	0	0
- electricity	8.666	8.666	7.910	7.910
Total	14.686	13.391	11.438	10.451

TABLE I.3 (cont'd)

	1974 actual	1980	1985	1990
3. Coke				
Industry	10.122	10.500	10.150	9.975
Transport	84	84	70	35
Small consumers	4.914	4.200	3.500	3.500
Transformation	5.285	6.160	9.310	10.500
Total	20.405	20.944	23.030	24.010

(c) Production Targets, Import Requirements and Export Policy

It is difficult for companies to plan the further development of coal production in Austria up till 1985. The reasons for this are the further course of substitution which is virtually unpredictable and the great uncertainty in judging the further development of costs and returns. Consequently the following production targets (brown coal only) are only indicative:

1976 (actual)	3.1 mill.t
1980	2.9 mill.t
1985	2.3 mill.t

Import requirements will be as follows (in mill.t):

	1976 (actual)	1980	1985
Hard coal	2.8	2.67	2.6
Coke	1.0	1.29	1.6
Brown coal and brown coal briquettes	0.65	0.83	0.8

The economic situation of the Austrian coal mining companies is unfavourable and constantly worsening, but, in principle, a solution to existing problems seems possible. Apart from acute structural problems, the cause of this unfavourable economic development is primarily the ever-widening gap between costs and returns. For prosperous further developments it would be important to secure the coal market, especially with regard to thermal power stations and long-distance heating stations.

The crisis of 1973/74 showed the advantage of domestic energy sources. Even though the importance of domestic coals is relatively modest, its influence in meeting supply shortages and alleviating the effects of increased prices of imported energy on the balance of payments should not be underestimated.

B. BELGIUM

In the preparation of forecasts, a clear distinction was made between coking coals, which alone are suitable for meeting the specific needs of the iron and steel industry - assuming that the traditional process of steel manufacture will continue to be used throughout the period in question - and "energy" coals, which have been gradually replaced by petroleum products and natural gas in the other sectors of consumption, and more particularly in various industries, in households and in electric power stations.

The rise in the price of fuel oil, which is responsible for the energy crisis and constitutes its most important aspect, has had no repercussions on sales of coking coal because of the specific use to which it is put.

On the other hand, a return to coal was noted in 1974 in connexion with electricity production wherever it was possible to reconvert multi-purpose units to coal.

It was also found that the secular recession of coal was only partially slowed in the case of household consumption.

No change in earlier trends appeared in deliveries to the various industries.

It was in the light of those factors, therefore, that this coal balance was drawn up and that Belgian coal policy was readjusted within the framework of the guidelines for energy policy drawn by the Commission of the European Communities.

There would no longer be any exports, except to certain traditional markets, involving a tonnage that seems of no importance in relation to the total balance.

Mention should also be made of the studies which have been undertaken with a view to underground gasification of coal under certain conditions. If this research is successful to the point where exploitation of the process would be practicable at the industrial level, it is obvious that this method of utilizing coal would make it possible to recover the large coal reserves which still exist in Belgium but whose exploitation is out of the question under normal conditions.

In the preparation of the following medium-term forecasts, account was taken of the following factors:

(a) Coal production in the Campine Basin (coking coal) would be maintained at an annual level of 7 mill.t, while in the Southern Basin (anthracite), because of gradual exhaustion of the deposits and increasingly high operating costs, measures were adopted to stop the granting of subsidies, thus bringing mining to a halt, so that the seven mine workings which are still in operation will be closed down by 1981.

(b) The current crisis in the iron and steel industry is not only cyclical but also structural. Consideration is therefore being given to a rationalization programme, implementation of which will probably be accompanied by the setting of a lower production target than was previously envisaged. Pending final determination of that target, Belgium's estimated coal balance for 1985 has been established on the assumption that crude steel output will amount to 13 mill.t.

(c) In order further to safeguard energy supplies, a certain number of electric power plants have already been reconverted to coal or will have been reconverted by 1980.

TABLE I.4 Tentative Coal Balance For 1985

10^6 tce

	1973	1974	1985
Sales			
Coking coal	9.9	10.5	8.6
Coal for electric power stations	2.1	2.6	6.3
Industrial coal	0.9	0.9	0.6
Household coal	3.3	3.2	0.6
Domestic sales	16.2	17.2	16.1
Exports	0.5	0.5	0.2
Total sales	16.7	17.7	16.3
Availabilities			
Production	8.2	7.6	6.5
Recovery	0.6	0.6	0.7
Imports	7.3	9.5	9.1
Total availabilities	16.1	17.7	16.3
Changes in stocks	-0.6	-	-

Furthermore, the electrical industry's equipment programme provides for the construction of another coal-fired plant with a capacity of 600 MW which would come on stream in 1983.

This expansion of installations operating on coal means that in 1985 coal consumption would amount to 6.3 mill.t as compared with 2.6 mill.t in 1974.

(d) The imports forecast for 1985, i.e. 9.1 mill.t, can be broken down among the various sectors of use as follows:

- coking coal	4,700,000 tce
- coal for electric power stations	3,800,000 tce
- industrial coal	-
- household coal	600,000 tce
	9,100,000 tce

C. BULGARIA

The problem of a further development of the national economy is closely linked with the creation of a modern, highly efficient energy economy. The development of the energy economy, and more particularly, of electric power generation provides the material base for the practical application of scientific and technological advances and for improvement of the people's working and living conditions.

The growth of energy consumption will be accompanied by significant changes in the composition of energy resources. The tendency for fuel oil, natural gas and atomic energy to cover an increasing share of energy consumption will develop even faster. This will be one of the factors in increasing the economic efficiency of the country's energy balance. The characteristic feature of the long-term energy balance will be a relatively high degree of electrification as electricity consumption outstrips the direct consumption of primary energy resources.

As a result of the development of material production, the rise in the people's level of living and the consequent changes in the public utility and service sector, the country's energy requirements are expected to increase by more than 100 per cent between 1975 and 1990. The mean annual rate of growth in energy consumption is expected to be 8.1 per cent in 1976-1980, 6.0 per cent in 1981-1985 and 4.2 per cent in 1986-1990. The proportion of the national income spent on energy is expected to decline accordingly.

The relatively high rate of growth in energy consumption expected in 1976-1980 will be brought about mainly by a great increase in production in the chemical industry, the metallurgical industry and other energy-intensive sectors.

The rapid growth of energy requirements and the tendency to raise the qualitative demands made on the process of energy supply create the problem of finding vast quantities of high-quality, high-efficiency energy resources over the period up to 1990.

The main local soucce of energy over the period up to 1990 will be coal, 92 per cent of which will be brown coal won by open-cast mining. In view of the poor quality of the local coal, brown coal will be used mainly for the generation of electric power (approximately 85 per cent by 1990) at large-scale electric power stations with high technical parameters.

In order to meet its energy requirements, Bulgaria will import fuel oil, natural gas, atomic fuel for electric power generation, coal and electric power. Imports of energy sources will be assured primarily by integrating the People's Republic of Bulgaria more thoroughly with the Soviet Union and other countries members of the Council for Mutual Economic Assistance (CMEA). In the process of solving the energy problem, the scope for co-operation on a mutually advantageous basis with other countries will also be explored. Deliveries of energy resources will be affected by expanding regular flows of trade, through participation in investment in the construction of power engineering installations in other countries, by linking the electric power systems of CMEA countries with those of other neighbouring countries and so forth.

In view of the great increase in the country's requirements of energy resources which is expected in the future, the intention is to develop the energy economy on the following main lines over the period up to 1990.

Fuller use of local energy resources in keeping with their economic efficiency. To this end, geological prospecting operations will be continued in search of new deposits of petroleum, of natural gas and of the raw material needed for atomic power engineering.

Construction of new hydro-power installations according to requirements for the peak capacity and for the multi-purpose use of water. Investigation of the scope for worthwhile use of geothermal energy and secondary energy resources, and for including new resources in the energy turnover (thermonuclear fusion, magnetohydrodynamic generators, etc.).

Intensified study of the efficiency of imported energy sources; development of broader and deeper relationships of integration with CMEA countries.

Further refinement of the sectoral composition of the national economy with a view to reducing the energy intensity of production.

Closer co-ordination of standards for energy consumption with the requirements of increased efficiency of social production on the basis of scientific and technological advances.

Relegation of conventional electric power engineering to a less dominant role in the energy economy as a whole, which will be based on a rapid increase in the generation of electric power at nuclear power plants.

D. CANADA

(a) Total Demand Forecast

The consolidated "primary energy" demand forecast for all sectors of the economy provides an estimate of the total energy which must be produced, at source, to supply Canada's requirements.

TABLE I.5 Forecast of Primary Energy Consumption

(Terajoules x 10^6)

	1970	%	1980	1990	2000	%
Petroleum	3.3	48	5.1-5.7	7.1-10.8	9.8-15.4	} 63
Natural gas	1.3	19	3.3-2.6	6.3- 2.6	8.7- 3.2	
Coal	0.7	10	1.2	1.7	2.0	7
Hydro-electricity*	1.6	23	2.4	3.3	3.6	12
Nuclear and other	-	-	0.5	2.1	5.4	18
TOTAL	6.9	100	12.5	20.5	29.5	100

Note: Figures may not add up to totals due to rounding.
* Hydro-electricity at 10.55 megajoules per kWh (10,000 BTU per kWh).

For this demand to be satisfied, oil production will have to increase to the range of 0.7 to 1.1 million cubic meters per day (4.4 to 6.9 million barrels), natural gas to 85 to 230 billion cubic meters per year (3.0 to 8.2 trillion cubic feet), coal to 90 mill.t per year. Because Canada has been a supplier of export markets for each of these commodities, however, these rates of production may be less difficult to achieve than would otherwise be the case. Continued supply of the export market may also remain a viable option.

(b) Coal Resources

Coal reserves if Canada south of 60° north latitude have been estimated at 109 bill. t in the ground, of which almost 98 per cent occur in the three western provinces. In terms of the degree of confidence in these estimates, some 8.5 per cent of the total coal reserve is classified as "measured", 42.2 per cent as "indicated" and the remainder, 49.3 per cent as "inferred". No accurate measure of economically recoverable coal is yet available. Excluded from the reserves estimates are deposits, such as those north of 60° latitude, for which there are insufficient data to permit estimation of resources. In terms of rank, it has been estimated that approximately three-quarters of Canada's coal reserves are low and medium volatile coal, with the remaining one-quarter of the reserves being approximately equally divided among high volatiles, sub-bituminous and lignite.

During the 1960s, Canadian coal production averaged about 10 mill.t per year. Demand is greatly increasing in the 1970s as evidenced by the 1972 production rate of 18.7 mill.t and the 1973 production rate of 20.5 mill.t. Thus, comparing Canada's recent coal production rate with measured coal reserves only, the Canadian coal has a reserve life index of some 500 years. On the basis of inferred reserves, the life index is measured in thousands of years.

The prairie mines of Saskatchewan and Alberta, which produce lignite and sub-bituminous coals from surface mining operations at a cost of $2-$4 per ton possess a substantial production capability and have the potential for a strong expansion. This potential arises not only from the large reserves of economically available coal, but also from the relatively good and reasonably predictable mining conditions. Productivities are high, being in the range of 35-45 t per man-shift. An expansion of production capability does not present unusual problems. The main market is the growing electric utility industry of the prairie provinces which consumed approximately 7 mill.t of lignite and sub-bituminous coals in 1972. The prediction is that this market will grow to about 13 mill.t per year in about 10 years. When and if gasification of coal becomes economically attractive in Canada, it will likely be based on these low-cost prairie lignite and sub-bituminous coals.

Further westward, the mountain coal fields of Alberta and British Columbia produce mainly metallurgical coals at a cost range of $13 to $20 per ton. Production capability in this area has substantially enlarged since 1967. The first objective of the operators is to reach by the mid-1970s, the total annual demand of their contracts with Japanese buyers of approximately 13 mill.t per year. Currently the total output of these mines is about 9.6 mill.t per year. The forecast is that they can reach a capability of 20 and perhaps 30 mill.t per year in the 1980s, depending upon the assurance of viable markets.

Canadian steam-raising coal is in competition with coal from the United States as well as with oil, natural gas, uranium and hydro-electric power. Its competitive position depends not only on mining conditions and production costs at the mine, but significantly also on distance between mine and market. The transportation cost is an important item in its selling price to consumers and is a major reason why western Canadian coals are not employed in any significant quantities in Ontario. This important central market has always depended heavily upon the nearer coal fields of the eastern United States. Obviously, as pressures increase in the United States for energy self-sufficiency, the availability in Canada of new supplies from United States sources will decrease. This factor, along with sharply rising oil and gas prices will improve the feasibility of supplying western Canadian coal to thermal power generation stations in Central Canada.

(c) Conclusion

The demand forecasts presented herein indicate that the total Canadian requirements of various energy commodities for the remainder of the century range as follows: crude oil 5-8 billion cubic meters (30-50 billion barrels) natural gas 2-4 trillion cubic meters (75-150 TCF), coal 2-3 billion t and uranium 150,000 t. While proved conventional oil and gas reserves are small in relation to these needs, less than one-half of the estimated undiscovered oil and gas potential could be sufficient to meet demand through the year 2000. Recoverable oilsands and heavy oils exceed the petroleum demands of this century by a multiple of 5 to 8 times. Similarly, proved coal reserves exceed cumulative demand to the year 2000 by a factor of 3 to 5 times. Reasonably assured uranium reserves are more than triple total anticipated demand to the year 2000, while potential reserves up to a price of $33/kg ($15/lb), would exceed this demand by a factor of about fifteen.

To meet domestic energy demands in the year 2000, production capability in the primary energy industries will have to be increased by a factor of about three for crude oil and natural gas, by a factor of about five for coal and about four for uranium. Hydro-electricity production must at least double. The capital investment required to find and develop the energy resources capable of satisfying Canadian demand is estimated in the range of $250 billion (constant dollars) through the year 2000. Downstream facilities for processing, distribution and marketing will require additional investment of like magnitude.

Coal reserves, while vast in magnitude, are located far from the large population centres of eastern Canada. High transportation costs and the relatively high cost of coal utilization, unfortunately, assure that coal burning for steam raising will be at best only competitive with other energy sources. Gasified or liquified coal based upon low cost prairie strip-mined feedstocks may provide large supplies of secure and environmentally acceptable energy. These products, however, will not be low in cost.

Over many years, Canada has imported substantial volumes of coal from the United States. These imports at near the present level are expected to continue. Over the last decade, exports of Canadian coal have increased to a point of 50 per cent of imports in 1973. It is likely that exports of Canadian coal will continue to increase.

E. CZECHOSLOVAKIA

Studies of research institutes on the share of the coal industry in the fuel and energy balance of the Czechoslovak Socialist Republic, prepared as part of the basic materials to be considered by planning bodies, make the following assumptions concerning the development of Czechoslovakia's coal industry.

1. The pattern of the long-term development and exploitation of Czechoslovakia's energy potential depends to a considerable extent on the specific structure and quality of its primary energy sources and on the complex conditions of recovery and utilization.

Economically speaking, the usable primary energy of water-courses and the proven reserves of bitumens (oil and natural gas) to the exploitation of which so much attention is given, represent a relatively unimportant part of primary energy sources and, in effect, only underline the non-complex nature of our fuel and energy base.

Consequently, in the next 10-15 years, the decisive role in the development of the fuel and energy base will continue to be played - along with the import of energy resources - by the mining and exploitation of our own coalfields which, however, are unlikely, on the basis of present knowledge, to reveal new deposits of sufficient richness to affect substantially the known situation as regards the geological reserves of coal.

The anticipated intensive working of these dispersed coal reserves in future years will, in conjunction with steadily deteriorating natural conditions of exploitation, give rise after 1985 to a slowing down in the rate of growth of new exploitable capacities. Within the overall fuel and energy balance this will manifest itself in a climax and thereafter a stagnation of solid fuel output around 1990.

2. The following trend (in per cent) is anticipated in the structure of primary sources:

TABLE I.6 Structure of Primary Energy Sources

	1970	1975	1980	1990
Total resources	100	100	100	100
Consisting of:				
Hard fuel	75.5	65.8	58.8	45.6
Liquid and gas fuel	20.9	30.3	36.5	41.2
Primary electrical and thermal energy	2.7	3.1	4.1	13.2
Other	0.9	0.8	0.6	-

By 1990 per capita need should reach 8.9 tce as compared with 6.2 tce in 1975, while the total internal demand on primary sources should increase by about 46-56 per cent between 1975 and 1990 (1975 = 100 per cent).

By 1990 imported fuel and energy should constitute approximately 46 per cent of total energy sources, as compared with 34.8 per cent in 1975.

3. The major part of our energy sources is accounted for by brown coal and hard coal. While more than half of the hard coal is used for coking, the bulk of our energy needs, especially for the generation of electric power, are met by brown coal and the remaining hard coal.

In the period up to 1990, mine output is expected to develop along the following lines:

TABLE I.7 Coal Output (mill.t)

	1970	1975	1980	1990
Hard coal	28.2	27.4	28.2	28
including coking coal	17.2	18.2	18.6	18-19
Brown coal and lignite	81.3	83.2	98	103-108

4. As far as consumption is concerned, there will be a steady expansion of the use of coal for energy - especially low-calorie brown coal - in the large electric power stations and district heating stations of the Ministry of Fuel and Energy and in other industrial sectors, which by 1990 should be using about 70 to 75 per cent of the total quantity of coal available for energy.

At present a review is being made of the technological and economic possibilities of constructing additional capacity for the gasification under pressure of brown coal from the country's main brown-coal producing area, the North Czechia coalfield.

As regards meeting the needs of the household sector, where traditionally graded coal has been widely used for furnace heating, it is expected that this fuel will first be gradually replaced by high-quality forms of energy - gas and electricity - and thereafter primarily by heat energy supplied from central district-heating sources. This system of supplying the household sector with heat energy is being, and will over the long term continue to be, introduced, especially in heavily populated urban districts.

In this way, by 1990 overall direct consumption of coal in the household sector is expected to be reduced by 50 per cent as compared to 1975.

Solution of the country's internal fuel-energy balance problem presupposes the continued imports of all types of fuel and energy, and share of natural gas and oil being increased.

As to exports, it is expected that the export of hard coal will be kept at its present level, that of brown coal slightly increased and that of coke gradually reduced.

F. FINLAND

(a) Coal

In 1974 the share of coal was about 9 per cent of the total energy requirements in Finland. The long-term target is, however, to raise the share of coal to the level of 11 to 12 per cent of our total energy demand in 1985.

The demand forecasts for coal are as follows:

TABLE I.8 Coal Demand Forecast

mill.t

	1974 (actual)	1985 (forecast)
Industry	1.3	1.4
Electricity generation thermal power plants	1.3	3.3
District heating plants	0.5	1.2
Households and traffic	0.1	0.0
Coke oven coke	0.8	1.9
Total consumption	4.0	7.8

There is no coal production in Finland and thus import requirements equal consumption.

About 88 per cent of the coal was imported from Poland last year.

Finland aims at raising the coal imports from Poland to the level of 6 mill.t per annum by the end of the 1970s.

(b) Peat

Peat is the only domestic source of energy which will witness a marked growth. Peat production will reach 20 mill. m^3. This is equivalent to about 2 mill.t of oil and about 5 per cent of the total anergy needs of the country in 1985.

Finland's peat reserves exceed 100,000 mill.m^3 and the average annual growth in her bogs is about 50 mill.m^3.

G. FRANCE

In view of the new situation created by the rise in price of petroleum products and of fuels in general, a fresh appraisal has been made of the French energy policy, based on the following main objectives:

- reducing energy costs in the medium term;
- rendering supplies less vulnerable.

In 1974, France had to rely on imports for 76 per cent of its total energy supply. With a view to achieving the above objectives, it has been decided to bring this figure down to between 55 and 60 per cent by 1985, through savings in consumption and the development of national production (new forms of energy; coal, oil exploration on the continental shelf and nuclear power).

(a) Energy Consumption

In seeking to reduce consumption by cutting down on waste and by investments, the aim is to limit energy consumption to 360 million tons coal equivalent in 1985 (instead of the 425 mill.tce forecast before the crisis).

France's natural resources in fossil fuels are very limited and the development of new forms of energy is not likely to have any great impact for several decades. In the present economic situation, nuclear energy is competitive [1] with a conventional thermal power station using coal or fuel oil costing 1.20-1.40 centime per thermie; but fuel oil in large quantities now costs around 3 c per thermie. Regardless of the uncertainty surrounding the cost of nuclear energy and developments in petroleum products, it became clear that not only would the whole increase in French electric power production have to be met by nuclear energy, but also that after 1980 the latter would advantageously replace a proportion of the conventional thermal energy supply, in other words it would replace fuel oil.

Accordingly, the present prospects are that there will be a large increase in nuclear production, which is expected to cover 25 per cent of French energy requirements by about 1985. The following table shows how the share of various forms of energy may develop in the next ten years.

TABLE I.9 Development and Structure of Energy Supplies

	1974		1980		1985	
	mill tce	%	mill tce	%	mill tce	%
- Petroleum	167.4	64	229	73	191	53
- Coal	46.9	18				
- Gas	24.1	9	39	12	55	15.5
- Primary energy	23.4	9	47	15	110	30.5
(including nuclear)	(4.6)	(2)	(26)	(8)	(907)	(25)
- New forms of energy					0.4	1
- Net electricity exports	-0.1	0				
TOTAL	261.7	100	315	100	360	100

As will be seen, the over-all demand for coal + petroleum products will show a moderate increase in absolute terms until about 1980, when it will start to decline as a result of the increasing production of nuclear energy.

As regards coal itself, it is preferable, regardless of any other considerations, to give it every opportunity, despite the meagre national resources.

1/ For base-load operation.

For this reason, it was decided to study how far and under what conditions it would be possible to stabilize French consumption up to 1985, at the 1973 level, i.e. at 30 mill.t of oil equivalent or 45 mill. tce. Pending the results of these studies, the following section contains an analysis of the coal demand forecast per sector of consumption.

(b) Coal Demand Forecasts

Each main consumption sector will be considered in turn.

Electric power stations. The expected coal consumption in electric power stations is, of course, closely linked to developments in electricity consumption trends (in TWh), will help to clarify the situation.

TABLE I.10 Electricity Consumption and Production

	1974 (actual)	1980	1985
Electricity consumption (in TWh)	179.7	275	385
Electricity production			
- primary energy (hydro + nuclear)	70.5	135	327
- conventional thermal			
coal + fuel oil 1/	85.9	116	34
other 2/	23.9	24	24
- less exports + pumped storage	-0.6		

1/ In EDF power stations, collieries and the iron and steel industry only.

2/ Gas, and also coal + fuel oil consumed by various independent producers.

It is probable that conventional thermal production will increase, though only moderately, in the period up to 1980, after which it will decline greatly as a result of the massive expansion of nuclear power.

It is also clear that estimates of these requirements will be very sensitive to the demand for electricity, since nuclear capacity will function as the base load capacity and, of course, to hydraulicity conditions.

The total capacity of EDF (Electricité de France) conventional thermal power-stations (existing or under construction) is 20,300 MW, including 6,200 MW from coal-fired or convertible power-stations. The latter capacity could be increased, if necessary, to 8,650 MW after reconversion to coal of equipment originally designed to be convertible.

In view of the prospective needs for conventional thermal power described above, there are, at present, no plans to construct new conventional thermal units - not for the EDF, at any rate. The existing capacity will probably be heavily utilized up to the 1980s (roughty 5,000 h annually), after which it will become surplus and some oil-fired power stations will probably have to be placed in reserve, conventional thermal power being used mostly to meet peak electricity loads, involving an average of less than 2,000 h annually.

EDF's convertible power stations will make it possible to increase the use of coal quite appreciably in the years immediately ahead, at least if the fuels on offer are competitive with fuel oil, having regard to the uses to which they can be put and subject to the balance of French petroleum refinery products.

On the other hand, the decline in global requirements after the 1980s will pose very serious problems if a too rapid fall in coal consumption is to be avoided. It is conceivable that priority might be given to using coal-fired power stations along with reconversion to coal of convertible equipment. But such operations are costly and would not prevent a very considerable drop in coal consumption.

Coal consumption in mine power stations, whose total capacity is 3,460 MW, is naturally linked to the availability of low-grade products and of mixed coals. It may even be necessary for them to have recourse to imported coal in the next few years if there is a strong revival of the economy; conversely, after the 1980s, the marketing of low-grade mining products might necessitate the construction of new electricity units, owing to the aging of equipment.

It should be mentioned that EDF is now working an open-cast brown-coal mine in the Landes (Arjuzanx). It is intended to expand production to the maximum during the next few years, and the reserves will be exhausted early in the next decade.

The prospects for the consumption of coal by electric power stations are summarized in the following table (in mill.t):

TABLE I.11 Coal Use in Power Stations

Power stations		1973 (actual)	1974 (actual)	1980	1985
EDF	hard coal	5.2	5.2	10/12	4.5/6.5
	brown coal	1.3	1.2	1.6	0.3 a/
MINES	hard coal and brown coal	7.8	7.5	5.7	3.5
Iron and steel industry		0.3	0.2	0.3	0.2
Total	in real Mt	14.6	14.1	17.6/19.6	8.5/10.5
	in TWh	28.2	28.0	36/41	19/24
Pre-crisis forecasts		-	-	approx.8	4

a/ Provence brown coal

It is pointed out that these are target projections and in the case of 1985, are based on the assumption that convertible power stations will be re-equipped for coal and that priority will be given to the use of coal-burning plant, with a view to attaining the higher of the two target figures given.

Coal consumption in industry. The French industry uses 1.6 mill.t of coke (mostly for specific non-substitutable needs) consumption of which is slowly declining, and 2.6 mill.t of coal (in 1974), which is directly exposed to competition from other fuels.

For the past ten years, consumption in this sector has been falling by 0.8 mill.t annually. The new energy situation makes it necessary not merely to stabilize, but to increase consumption in this sector, substituting coal for fuel oil, particularly in the case of cement works. A study is being carried out to this end, but a reversal of the industry's trend comes up against the problem that it implies the abandonment of modern oil-fired installations which would have to be replaced by more costly coal-burning equipment.

In connexion with possible recourse to imports, it should be noted that the monthly volume of industrial consumption is disproportionate to the capacity of coal-ships and there are very costly problems of storage and distribution.

In any event, such an operation would clearly be impossible without stable supplies, in terms of quantity, quality, and a price that remains appreciably lower than that of fuel oil.

Depending on the scope of eventual specific decisions on the matter of capital equipment as a result of these studies, and on the effectiveness of efforts to promote coal consumption, it is estimated that consumption in this sector will amount to 4-7.2 mill.t in 1980 and 3.5-10 mill.t in 1985.

Residential and tertiary sector. Consumption in this sector has been declining steadily over the past ten years, at the rate of approximately one mill.t annually. The oil crisis has retarded conversions of equipment by a year, but these have now been resumed, despite the rationing of fuel oil and attractive coal prices. It was assumed that the rate of decline would be arrested, but the decline now appears to be inevitable owing to the inconveniences involved in the use of coal. A bracket of 3-4.5 mill.t has been adopted for 1985, as compared with the 8.3 mill.t in 1974.

Iron and steel industry. It has been assumed that total coke and sintering-coal requirements will stabilize at around the present level, so that the following are the estimates for 1985 (in mill.t):

- Coke (blast furnace and other)	12.8
- Hard coal for sintering	2.7
	15.5

Miscellaneous and recapitulation. Consumption under this heading includes the mines' own requirements and processing losses during coking operations, deliveries being counted at the stage of the end consumer in the form of coke, where the fuel is delivered in this form. Since French coal resources are inadequate for the country's needs, there are no plans at present to expand the market by converting coal to oils or gas.

Expected coal requirements of the various sectors are recapitulated in the following table:

TABLE I.12 Forecast Trend of Coal Deliveries

(in mill.t of hard coal + brown coal + coke + briquettes)

Consumption sector	1974 (actual)	1980 (forecast)	1985 (forecast)
Iron and steel industry	15.7	15.5	15.5
Industry	4.2	4-7.2	3.5-10
Residential and tertiary	8.3	5-6	3-4.5
Producers	0.5	0.5	0.5
Processing losses	3.8	4.4	4.5
Thermal power stations	14.3	17-19	8.5-10.5
Total	46.8	46-52.6	35.5-45.5

N.B.: The higher of the double forecasts are based on the assumption that total coal consumption will be maintained by investment in the sectors concerned and by effective measures to stimulate consumption. They do not imply, therefore, that the market has been created. Mobilization of resources for each sector can only be envisaged if there is a corresponding consumer demand.

Lastly, it is interesting to note that the above forecasts for the period up to 1985 are 10-22 mill.t above those made before the oil crisis (25 mill.t).

(c) Production, Imports and Exports

Following the upheavals on the energy market at the end of 1973, a fresh inventory was made of national coal resources economically exploitable in the light of the prices of imported fuels. This review led to the preparation of a new coal production plan, approved in 1974, which reflects a slowing down in the rate of decline of extraction, from roughly 3 to 1 mill.t annually.

But this plan is not rigidly tied to the programme prepared in 1974 and may be adapted, provided CdF (Charbonnages de France) observes the criteria of economic viability adopted. Thus, to take into account changes in the situation and, in particular, recruitment possibilities, additional projects have been drawn up. On 6 November 1975 the Government took three important measures of developing production in Lorraine, concerning the mines of Merlebach, Simon-Wendel and Sainte Fontaine.

TABLE I.13 Latest Decisions on the Development Perspectives of Coal Production

(in Mill.5)

	1974	1980	1985
Nord-Pas-de-Calais	9.0	4.8	0.7-0.8
Lorraine	9.1	10.5-11.9	10.5
Centre-Midi	6.4	5.1	2.7-3.6
Sub-total	24.5	20.4-20.9	12.4-14.9
Recovery products	1.1	0.6	0.5
Arjuzanx brown-coal	1.2	1.6	-
Total	26.8	22.6-23.1	12.9-15.4

As will be seen from the above table, output in the Nord-Pas-de-Calais coalfield is likely to decline owing to the exhaustion of deposits. On the other hand, efforts will be undertaken in Lorraine, whose considerable reserves could be developed, at certain collieries, under economically acceptable conditions.

These production prospects taken together with the above estimates of market possibilities at home and with export forecasts (which are very low) make possible an assessment of the magnitude of potential imports if the coals offered are competitive with fuel oil.

TABLE I.14 Coal Balances

(in mill.t)

	1974	1980	1985
National production	26.8	22.6-23.1	12.9-15.4
Imports	21.3	24.9-31.0	24.1-31.6
- Exports	- 1.8	- 1.5	- 1.5
+ Withdrawn from stocks	+ 0.5		
Net resource adjusted to the level of requirements	46.8	46-52.6	35.5-45.5

This table confirms the upward trend of imports, at least until 1980.

Imports in 1974 can be broken down into the following broad categories (in mill.t):

coke and coking coal	13.47
coal for electric power stations	3.54
other coals 1/	4.29
	21.30

With regard to the future, a basic trend is noticeable towards increased coking coal requirements as well as lean-coal requirements for ore-sintering. This increase will be limited in time by steam-coal requirements for EDF, with a possibly sharp decline as from the beginning of the next decade.

Lastly, expansion of the use of imported coals in industry and in households will not be possible unless the coals are offered at prices that will stay below the level of fuel-oil prices, on account of investment costs for boiler-houses and of problems connected with convenience of use (for households).

H. GERMAN DEMOCRATIC REPUBLIC

(a) Main Trends of Power Use and Raw Materials Problems in the Period Until 1990

The guaranteed long-term supply of the German Democratic Republic's national economy with raw materials and power is a decisive prerequisite for the further continuous dynamic development and the efficient structure of the process of reproduction of the national economy. Under the natural conditions of the country this is connected with high expenditures of the national economy.

On the basis of the joint long-term planning of the CMEA member countries with stable physical and value relations the following objective conditions and prerequisites have to be taken into account for the national economy:

- the mining of important domestic raw materials proceeds under more and more unbeneficial geological conditions;
- the extensive increase of the production is possible only for few raw materials;
- the world market prices for raw materials underlie a dynamic whose long-term trends and levels cannot be assessed with certainty at present.

Within an average annual growth rate of the national income of about 5 per cent and of the industrial commodities production of 6 to 7 per cent the aim is to manage with a growth rate of the raw materials production that is much lower than that of the industrial commodities production, and to decrease the specific consumption annually for all-important raw and synthetic materials.

(b) Specific Problems in the Field of Energy

Crude lignite is the only important domestic source of primary power.

1/ Including coals for ore-sintering.

With an average calorific value of 2,000 kcal/kg its utilization during the period till 1990 and beyond that will mainly be determined by:

- generation of electric power in power plants with large block units;
- coupled generation of heat and electric power in industry and electricity power and heat stations;
- generation of heat and hot water in electricity power and heat stations and a large number of local boiler plants;
- briquetting and further processing of briquettes to lignite high-temperature coke, low-temperature coke and town gas.

As the geological and hydrological mining conditions in the open workings become more and more difficult, the problem is to balance the increasing high costs through measures in the field of science and technology. In particular this is also necessary, because the increasing amount of overburden to be uncovered to an ever greater extent requires the stable continuation of the alraedy high level of production, the more as we are facing the problem of substituting permanently the capacities of excavated open workings, reserves of the single fields being limited.

In general the utilization of crude lignite represents a dominating factor for the balance of primary power; however, we have to put up with significant transformation losses as a result of the necessary number of transformation stages for the supply with final electric power, and based on the low energy efficiency.

For these reasons and because of the much faster development of the demand of electric power, especially in the non-producing sphere, which anyhow is connected with high transformation losses, we have to reckon for the long term with a growth rate of primary power that amounts to 50 per cent of that of the industrial commodities production. Thereby electric power again increases faster than the other sources of final power, so that the growth rates of primary power are also higher than those of the final power altogether.

A more intensive and efficient utilization of this raw material requires scientific-technicological investigations and solutions regarding the following:

- rationalization of the removal of overburden;
- improvement of the device for overburden removal;
- new methods of utilization and mining of deposits without open lode in the traditional way;
- methods for the utilization of lignite deposits with a specific structure, which cannot be worked at present because of high costs;
- methods for the more efficient prevention of the emission of harmful substances and better utilization of by-products (ash).

The demand of the national economy of oil and gas with a high methane content can be covered only by imports. In the national economy they are used purposefully under the aspect of the utilization of their properties as a source of power with a high density of heat and good adaptability to processing conditions in consuming spheres, where it is not possible to achieve these results otherwise with acceptable economic and technical expenditures.

Thus, the demand of both sources of power is mainly determined by substance-processing methods of the chemical industry, technological high-temperature-processes in metallurgy, glass and ceramic industries and others, and by transportation processes.

At the same time the stage of development of power generation for the national economy by this means is adapted step by step to the minimum requirements of technology and the demand of the population, and an improvement of the structure of primary power is achieved.

The emphasis of scientific-technological efforts in this field is on the development of methods and processes, which alow the maximum utilization of the raw materials oil/gas (production of aromatic hydrocarbons, plasmachemical processes, etc.).

Nuclear power base is being developed according to plan, and its share in electric power generation is increasing without affecting the dominating position of thermal power plants based on crude lignite, until 1990.

Within the framework of the CMEA we are participating in scientific-technological research work, which in particular aims at the improvement of nuclear power plants with thermal reactors of large block units and at the improvement of parameters and fast breeders.

Ensuring the overall energetic basis for the national economy in all consuming spheres requires the steady decrease of power consumption for specific purposes, and the development of techniques economizing power. For this reason industry, transport and construction have got special duties (e.g. thermal insulation). The use of interruptible contracts for electricity supply which has to be investigated further, among others, is one important measure of rational power consumption and of mastering the peak load problem. The trend of decreasing specific power consumption will continue in the long term.

I. GERMANY, FEDERAL REPUBLIC OF

(a) Hard Coal

The situation before the energy crisis. In its first energy policy programme, presented in 1973, a global strategy was developed for influencing the future shares of different energy forms in total energy supply; one of the aims was to give hard coal its proper place in supplies. At that time and in the light of available markets it was believed necessary to continue reducing coal output to a level of 83 mill.t in 1978; this would have meant a decline in the share of coal in primary energy consumption from about 24 per cent in 1972 to 11 per cent in 1980.

Results of the energy crisis. Developments since October 1973 have shown that the world energy market has been revolutionized. The period of relatively cheap and unhampered supplies of petroleum, which accounted in 1973 for about 55 per cent of the Federal Republic of Germany's total energy consumption, now belongs to history, at least for the foreseeable future.

As a result, the Government was obliged to revise its first energy policy programme in order to take account of the changed situation. It started from the position that the coal reserves present in the country should be utilized to an optimum degree in meeting energy needs. In the first revision of the energy policy programme, in October 1974, the following measures were set out with this objective in mind:

- a stabilization of production capacity at the present level of 94 to 95 mill.t (=100 mill t according to ECE definitions) in the years up to 1980;
- increased support for investment by the mining industry;
- increased support for coal research, both in the fields of mining technology and with a view to new applications (coal liquefaction and gasification);
- an extension of stockpiling policy through the building of coal reserves amounting to about 10 mill.t;
- measures to secure sales at prices sufficient to cover production costs, in particular statutory regulations to ensure that hard coal is given an adequate share in electricity generation.

Sales in individual sectors. The uncertainties brought about by the instability of the world energy market as a whole represent one of the reasons why estimating sales in individual sectors is even more difficult than in the past. On the basis of present knowledge, it is assumed that there will be markets for a total of 90 mill.t of German hard coal in 1980. This figure divides up as follows (figures for 1973 in brackets).

Electricity industry	35 mill.t (31.4)
Steel industry	25 mill.t (25.8)
Home use and other industries	11 mill.t (16.0)
Exports	18 mill.t (27.3)

Taking imports into consideration, the share of hard coal in the overall primary energy consumption in 1980 is likely to be 82 mill.tce or 17 per cent. By 1985 this will probably fall to 79 mill. tce or 14 per cent. Imports, which in recent years have been between 8 and 9 mill.t annually, may rise in the years to come since it is planned under certain circumstances to increase the 6.6 mill.t annual quota for imports from outside the European Economic Community by 3 mill.t of coking coal imports to supply the steel industry.

Longer-term forecasts. In revising its energy policy programme the Government deliberately avoided making long-term forecasts for the period after 1985. Given present knowledge, it is not possible to quantify changes from the mid-1980s onwards because they depend on widely varying factors, some of them lying outside the energy sector, such as macroeconomic developments nationally and internationally, patterns of consumer behaviour and so on. A major determinant will also be the success of worldwide efforts to develop alternative energy sources and new energy technologies. It can be expected that gasification of coal will not become a viable commercial proposition until the mid-1980s and after. In terms of quantity, the effects will not be properly felt much before 1990.

On the other hand, since the assessment of long-term developments is of considerable significance in an energy policy context the Government has commissioned several academic studies covering energy market prospects up to the year 2000; results are expected in the course of this year.

(b) Lignite

Lignite or brown coal is a particularly cheap form of energy and is definitely being utilized in the best way possible at the moment. The majority of output is used to generate electricity (83 per cent) while the remainder is converted into briquettes for use as a household fuel. The opening up of new open-cast sites will ensure that roughly the present level of output (1974: about 126 mill.t) will be available in 1985 and 1990.

(c) Hard coal and lignite

A further updating of th- energy programme is presently undertaken by the Federal Government. On 25 March 1977 it published (Bulletin, no. 30/77, p. 265ff) the principles and basic data of its energy programme thereby confirming its earlier decision to maintain production capacities at 94-95 mill t (100 mill t according to ECE definitions). This publication contains *inter alia* the following forecast of primary energy demand for 1980-85 elaborated by economic research institutes.

TABLE I.15 Primary Energy Consumption

	1973		1980		1985	
	mill.tce	%	mill.tce	%	mill.tce	%
Petroleum	208.9	55	216	50	226	45
Hard coal	84.2	22	72	17	73	15
Natural gas	38.6	10	73	17	87	18
Lignite	33.1	9	35	8	35	7
Nuclear power	3.9	1	28	6	62	13
Other	9.8	3	11	2	13	2
Primary energy consumption	378.5	100	435	100	496	100

J. GREECE

According to the energy policy applied by the Greek authorities, solid domestic fuels (lignite) are used for the production of electric power.

The aim of the Public Power Corporation is to exploit the identified lignite reserves of Ptolemais (1,080 mill.t) and of Megalopolis (470 mill.t), whose net calorific value is, respectively, of 1320 kcal/kg and 1000 kcal/kg, to produce electric power.

Thus, whereas in 1974 the consumption of lignite was 6.6 mill.t from Ptolemais, 4.4 mill.t from Megalopolis and 0.5 mill.t from Aliverion, it is anticipated that in 1985 it will amount to 25.4 mill.t from Ptolemais and 15.6 mill.t from Megalopolis.

It should be noted, that the anticipated highest potential of electric power which will be produced from lignite, during the period 1980-1990, is estimated at 29,000 GWH.

Furthermore, 1.5 mill.t of lignite are used for the production of nitric fertilizers.

K. HUNGARY

Planning directives make it reasonable to anticipate that Hungary's gross national product will rise by 5 to 6 per cent by 1990. The increasing rate of growth of the coal mining industry as one of the leading industries from the standpoint of the economy and energy-intensiveness, reflects, first and foremost, greater use of domestic mineral resource.

International experience confirms that consumption of basic fuels is increasing rapidly throughout the world as a result of two opposing trends - automation and mechanization being factors that tend to increase consumption, while the improved energy efficiency of machinery and changes in the structure of primary energy consumption tend to reduce it.

A distinctive feature of the increasing energy consumption pattern in Hungary over the next 15 years is that a larger porportion of energy will be consumed in the form of electric power. The rate of growth of electric power consumption is outstripping the rate of total energy consumption (over the long term the annual increase is expected to be approximately 7 per cent) by about 60-80 per cent. This trend is the inevitable result of the increasing use being made of technological equipment to improve productivity and efficiency and to raise levels of living, i.e. the rapid rise in population's consumption level. During the period 1980 to 1990, something of the order of 50 per cent of the total increase in energy consumption will be accounted for by the increased use of electric power.

The structure of energy sources is determined in advance by directives governing energy development. The relative share of hydrocarbons will continue to grow until 1980, reaching 64-65 per cent as against 42 per cent in 1970. In the years after 1980, as a consequence of the construction of coal-fired electric power stations the growing role of nuclear energy and better management, the share of hydrocarbons will remain at the same high level, gradually falling to 62-63 per cent by 1990. The relative share of solid mineral fuels will continue to decrease, and in 1990 will amount to approximately 21-22 per cent as against 52 per cent in 1970. Conversely, the proportion of nuclear energy, imported electric power and other energy sources will increase from 6 per cent in 1970 to about 16 per cent in 1990, nuclear energy accounting for approximately 9 per cent of the latter figure.

Even with maximum utilization of domestic energy resources, the Hungarian energy economy will become increasingly dependent on energy imports. The proportion of imports, including nuclear raw materials, will increase from 37 per cent in 1970 to 67 per cent by 1990. Owing to the increasing role of fuel imports, the extent to which realistic long-term development projections can be made will, to a large extent, depend on the sources of the country's growing import requirements and the conditions under which they can be satisfied.

Compared with other members of CMEA, Hungary's energy production is more costly and its import requirements greater than the average. These drawbacks are only partly offset by Hungary's relatively lower energy requirements and somewhat more favourable energy structure. These objective conditions make it necessary to give a great deal of attention to expanding co-operation on energy matters and, particularly, to developing links with the Soviet Union.

TABLE I.16 Structure of Energy Sources

	Unit of measurement: mill.tce		
	1970	1980	1990
Coal	16	12-13	15-16
Hydrocarbons	13	30-31	44-46
Nuclear energy	-	-	7
Others	2	4	5
Total:	31	46-48	71-74
	Unit of measurement: per cent		
	1970	1980	1990
Coal	52	26-27	21-22
Hydrocarbons	42	65	62
Nuclear energy	-	-	9-10
Others	6	8-9	7
Total:	100	100	100

L. IRELAND

(a) The Role of Coal

Coal represents about 9 per cent of Ireland's consumption of primary energy.

(b) Demand Forecasts

No worthwhile demand forecasts are possible for the years 1985 and 1990. The known indigenous resources are not sufficient to permit synthetic oil or gas manufacture and if electricity generation from coal is decided upon then additional sources of supply will be required.

(c) Production Targets

In the absence of coal-mining industry in the accepted sense, no production targets can be quoted and consequently, export policies do not arise. Import requirements of about 700,000 t per annum are expected to continue at that level unless electricity generation from coal becomes a feature of the industry.

M. NETHERLANDS

A new element of policy is the deliberate limitation of the growth of energy consumption. After the experiences following the oil crisis this has become a more permanent policy.

The traditional role of the Government in energy administration will have to increase.

Until some time ago the objective of energy strategy was the guaranteed, uninterrupted supply of every demand for energy at the lowest possible cost, with the choice left to the users. This objective was in conformity with the well-known five socio-economic macro-objectives of government policy: balanced growth, full employment, retention of money values, a balance of payments surplus and fair income distribution.

In the meantime, factors such as the awareness of the finite nature of basic material reserves, the greatly increased attention to our environment, and the recent oil crisis, have compelled further reflection on the objectives of our energy policy. Consequently, the objective now becomes "A guaranteed supply of the demand for energy, resulting from the most economical consumption possible, at the least possible social cost; supply and demand must fit into a more balanced system of international labour and income distribution and must be weighed against the needs imposed by ecological adaptability, safety and the employment situation".

For the time being, a number of variants have been formulated within the framework of energy policy in the narrower sense - policy measures or autonomous developments - and where possible the quantitative effect of each is given. The paper discusses in succession, without having the pretext of giving a complete enumeration:

- a slowing down of the growth in energy consumption by industry (by more economical consumption, a different pattern of industries, etc.);
- a slowing down of the growth in electricity consumption in the home and by public bodies, including active (official) information and by introducing an "economy test" for electrical equipment;
- slower growth in fuel consumption in the "household and public" sector, for instance by a movement in vehicles towards diesels, continuance of the "3 x 5" campaign and a different tax approach to petrol driven cars exceeding 2000 cc.
- operation in or about 1985 of three nuclear fission power plants of 1000 MW each;
- promotion of coal consumption by electricity generating stations to 3 to 4 mill.t a year; this may not reduce total energy consumption but it does improve the distribution over energy sources;
- changing building regulations for new houses so that all houses built after 1975 will have heat insulation, such as double glazing, roof insulation and filled-in cavity walls;
- measures to promote better insulation of existing homes having central heating;
- ensuring that the average temperature (as far as reached by heating) is reduced.

The overall effect of these measures would mean that up to 1980 the annual growth in energy consumption would be 4.5 per cent (against 5.3 per cent with a "no-change policy") and 2.2. per cent from 1980 to 1985 (against 4.2 per cent with no change).

Other features of this policy approach, which is a definite change as regards growth in energy consumption, are a much lower petroleum consumption and a rather smaller natural gas consumption. Nevertheless, the proportion of petroleum continues to rise (1973: 45 per cent, 1980: 50 per cent and 1985: 60 per cent) and there is barely any scope for forming a strategic natural gas reserve.

In 1950, coal was the main pillar (70 per cent) of Dutch power supplies. The situation has drastically changed in the meantime. At present only 4 per cent of the national energy requirements are covered by coal.

Coal production dropped from about 12 mill.t in the 1950s to less than 2 mill.t in 1973. Production ceased entirely at the end of 1974. In this connexion the major contribution by workers in the Limburg coal-mines to post-war recovery and hence to national economic progress had been stressed.

This country has technically recoverable coal reserves of 1,500 mill.t. But at present, the recovery of these mines is impossible. It is foreseen to appoint a commission to examine the long-term possibilities of mining Dutch coal. The expected trends in coal technology will have to be taken into account. In order that the operation of the Emma State Mine - now liquidated - should not be made completely impossible in future a number of conservation measures are being taken.

It must also be examined whether the Netherlands might play a part in operating the capacious world coal supplies, either for its own consumption or as a link in the international chain of trading and conversion. These possibilities gain emphasis in view of modern coal technology. Collaboration with countries experienced in coal conversion is an obvious step.

Policy is being aimed at promoting coal consumption in the Netherlands as much as possible - especially by electricity generating stations - assuming that the effect on the environment of coal burning can be kept within acceptable limits in properly equipped plants.

N. NORWAY

Parliamentary report No.100 (1973-1974), Norway's energy situation for the future, from May 1974, presents forecasts for energy consumption from 1975 to 1985. The report preassumes an annual growth in GDP of 4.5 per cent from 1974 to 1985 and in energy consumption of 3.4 per cent from 1974 to 1980 and 3.3 per cent from 1980 to 1985. The high oil price will probably result in a decreased oil consumption compared to hydroelectric power. The prognoses are presented in the table.

TABLE I.17 Energy Consumption 1974-1985
(million tons oil equivalents)

	1974	1980	1985
Total energy consumption	19.0	23.2	27.3
solid fuels	1.2	1.2	1.2
electricity	10.6	14.1	16.8
oil and gas	7.2	7.9	9.3
Source: Parliamentary report No.100 (1973-1974)			

The only Norwegian coal mine is operating on Svalbard (Spitzbergen), and today produces about 400,000 t for sale. Only a small part of this is exported.

The company is planning by 1985 to produce about 1.2 mill.t for sale, of which about 50 per cent will be exported. This also applies to 1990.

Because of the petroleum deposits found on the Norwegian continental shelf, one is likely to believe that new markets for coal, including electricity generation, synthetic oil and gas, etc. will not occur in Norway in the near future.

O. POLAND

The general pattern of Poland's fuel and energy economy is influenced by the following factors:

available sources of energy

the Government's policy aimed at making maximum use of domestically available raw materials, and

fuel price fluctuations on world markets.

Poland possesses the following natural energy reserves: extensive hard coal deposits in favourably situated formations, brown coal deposits which are not yet fully exploited, limited reserves of natural gas, and very small amounts of oil and uranium ore; there is no hope of increasing the share of hydro power in total electricity production.

In view of the above mix of energy resources, production of primary sources of energy has been expanded in Poland mainly through the development of the hard and brown coalmining industry. This is reflected in the structure of Poland's fuel and energy balance and in the increasing amounts of hard coal and coke exported every year.

The basic indicators showing the development of the hard and brown coalmining industry during the past 25 years are as follows:

TABLE I.18 Coal Production and Exports (mill.t)

Indicator	1950	1960	1970	1975
Total output of hard coal	78.0	104.4	140.1	171.5
of which:				
type 34-38 coking coal	5.9*/	8.3	16.9	25.0
of which: type 35.1	0.5*/	0.8	6.8	13.6
Output of brown coal	4.8	9.3	32.8	39.6
Coke production	6.0	11.9	16.5	18.4
Exports:				
- hard coal	26.6	17.5	28.8	41.3
- coke	2.3	2.1	2.3	3.0

*/ At 1951 prices.

This rapid expansion in hard and brown coal production and the relatively low cost of this fuel, as well as a slight increase in natural gas production and an insignificant amount of oil production, have all influenced the structure of Poland's fuel and energy balance.

The extent to which our country's energy requirements have been met with hard and brown coal is as follows:

1950	1960	1970	1975
95 %	90 %	82.3%	78 %

Hard and brown coal is the basic fuel used for the generation of electric power and heat in Poland, and in 1975 it accounted for 93.7 per cent of the country's electricity production. These two fuels are also of basic importance in meeting the energy requirements of other industries.

Reliance on hard and brown coal is gradually being reduced in those sectors of the national economy where its use is not justified by technological and economic considerations. These sectors include road and rail transport, ferrous and non-ferrous metallurgy and agricultural mechanization, as well as technological processes in the chemical industry.

A new branch of activity in which very large amounts of brown and hard coal are to be used embraces the chemical conversion of these primary sources of energy into manufactured gas of high calorific value, liquid fuels and solid smokeless fuels. For this purpose, a special Government programme has already been drawn up in Poland covering scientific and technical work in this field, as well as the introduction of the process itself.

Hard coal is to be used for the production of manufactured gas as early as 1979-1980.

The fact that the country's energy requirements are increasing at an annual rate of about 4 per cent in terms of standard fuel equivalent, the existence of favourable market conditions for sources of energy and new prospects for the chemical processing of coal (referred to above) all point to the desirability and need, under conditions prevailing in Poland, for the rapid expansion of the coalmining industry, not only in the next few years but also over the longer term, as part of the general development of the country as a whole. For this reason, economic plans covering the forthcoming 5-year periods as well as the 5-year horizon up to 1990 call for annual increases of 6 to 8 mill.t in hard coal production.

As part of the future development of the hard coalmining industry, existing mines are to be expanded and new ones opened up in the Silesian coalfield to increase the output of power station coal. Production of hard coal from a new deposit in the Lublin coalfield is also to be developed. The construction of a new experimental mine has begun at this deposit, and a considerable proportion of the anticipated increase in hard coal production will eventually be obtained from this coalfield.

Hard and brown coal in Poland, a country rich in this source of energy, will play a basic role in Poland's fuel and energy balance in the next few years as well as over the longer term. Processing industries using hard and brown coal as a feedstock are being established and steadily expanded; they include electric power stations, combined heat and electricity generating plants, heat generating plants and coking plants, all of which supply the consumer with high-quality forms of energy. According to the programme that has been drawn up, the use of coal for the production of manufactured gas of high calorific value, liquid fuels and solid smokeless fuels will be expanded in the near future. Our plans for the production of hard, coking and power station coal also call for the output of appropriate amounts for export assuming, of course, participation by the countries which intend to import our coal, i.e. their participation in the construction of additional production capacity.

Brown coal, which is a cheap fuel for the production of electric energy, is playing and will continue to play an important role in meeting fuel and energy requirements.

The electricity produced by power stations fired by brown coal accounts for 30 per cent of total electric power production.

A fuel and electric power combine is being built in the immediate vicinity of the Belchatow brown coal deposits. In the first stage of development, this deposit is expected to produce about 40 mill.t of brown coal per year. The aggregate brown coal output of this deposit will provide fuel for two thermal power stations with a total capacity of over 5,000 MW and an annual output of 32 billion kWh. Other brown coal deposits situated in our territory are to be developed in the future.

Possibilities of using brown coal not only for the production of electricity but also for chemical processing and the production of gas of a high calorific value are also being explored.

P. ROMANIA

As part of Romania's consistent efforts to ensure the steady and balanced expansion of the national economy, particular attention is being paid to the development of domestic primary energy resources, with a view to meeting its steadily increasing fuel and energy requirements, a policy which prevented it from being caught unawares by the world energy crisis in 1973.

In addition, the need for the more judicious use of hydrocarbons in connexion with the development of the petrochemical industry has led to increased use of solid fuels for the generation of electric power.

In these circumstances, coal production rose from 8 mill.t in 1960 to 20.5 mill.t in 1970; it is expected to reach 56 mill.t in 1980 and 72-77.8 mill.t by 1990.

The proportion of electric energy generated through the use of solid fuels has increased from 21.2 per cent in 1960 to 27.7 per cent in 1975; the corresponding figure for 1980 is expected to be 43.7 per cent.

The distribution of coal consumption in 1974 was as follows:

Iron and steel industry	5.6 per cent
Electric energy	73.9 per cent
Transport	3.2 per cent
Households	6.4 per cent
Miscellaneous	10.9 per cent

The expansion of electricity generation through the use of solid fuels is assured by the increase in lignite production, which is expected to rise from 14.1 mill.t in 1970 to 46.5 mill.t in 1980 and 60-65 mill.t by 1990.

The generation of increased amounts of electric energy through the use of solid fuels will be achieved, during the period 1976-1980, by the installation of 5,400-5,740 MW additional capacity in thermal electric power stations with high-capacity generators.

Special attention is also being paid to the exploitation of oil shale reserves in order to provide feedstock for thermal electric power stations and with a view to the complex utilization of oil shale ash.

In order to meet coal requirements during the 1976-1980 period, new mines with a production capacity of approximately 27 mill.t of lignite will be brought into operation. Of that amount, 72 per cent will be derived from open-cast mines.

The increase in lignite production up to 1985 will be assured by new mines already operational or still being developed and, during the period 1985-1990 new production capacity will be accounted for by underground mines.

In order to achieve competitive economic efficiency, lignite production will be concentrated in mines of progressively increasing capacity, as shown below:

	1970	1980
Underground mines	574,000 t per annum	985,000 t per annum
Open-cast mines	125,000 t per annum	5 mill.t per annum

Expansion of mining production will be achieved mainly by:

- greater use of highly mechanized and automated large-capacity equipment, in the case of open-cast mines;
- the increased mechanization and automation of winning, transport and extraction operations in the case of underground mines.

As from 1978, a deposit of oil shale with a calorific value of approximately 1,000 kcal/kg is to be exploited; output will be of the order of 5 mill.t in 1980 and 12-12.6 mill.t by 1990.

Coking coal reserves are, generally speaking, scanty and of relatively inferior quality.

Geological prospecting and prospecting in general is to be intensified with a view to the discovery of new deposits and the exploitation of existing deposits in a more efficient manner.

In accordance with these programmes, coking coal production will increase from 0.8 mill.t in 1970 to 2.4 mill.t in 1980 and 3.3 mill.t by 1990. However, these amounts will not be sufficient to meet all the requirements of industrial consumers, and Romania will still have to rely on imports in order to make up the shortfall.

Q. SPAIN

(a) Definition of the Role of Coal in the Total Energy Context

In 1973 the gross domestic consumption of primary energy was 84.3 mill. tce, of which coal accounted for 14.5 mill. tce or 17.1 per cent.

In 1980 coal consumption is expected to remain at a similar level: 17.9 per cent of the 130.7 mill. tce which will constitute gross domestic demand. In other words, total coal consumption will increase to 23.4 mill. tce.

Finally, in 1985, coal will contribute 25.5 mill. tce towards meeting total primary energy demand; in other words, coal will account for 14.7 per cent of gross domestic consumption.

A distinction should be drawn between the electricity sub-sector and the energy sector as a whole.

As far as the electricity sub-sector is concerned, coal is expected to constitute 36.5 per cent of primary energy consumption in 1980, and 45.0 per cent in 1985.

Coal will account for 37.0 TWh or 27.0 per cent of total electricity production in 1980, and 36.0 TWh or only 16.5 per cent in 1985.

As far as other sectors of energy consumption are concerned, coal is irreplaceable in the iron and steel industry, just as various petroleum products are irreplaceable. Coal consumption by the iron and steel industry is expected to amount to 9 mill. tce in 1980 and to increase to 12 mill. tce by 1985.

In addition to the coal consumed in the iron and steel industry and in thermal electric power stations, coal will still be used for other purposes in amounts of approximately 2 mill. tce per annum for the period 1980-1985.

(b) Demand Forecasts by Major Sectors and Qualities, as well as Greater and New Markets for Coal, including Electricity Generation, Synthetic Oil and Gas, etc.

The following three major sectors are taken into account in demand forecasts: the iron and steel industry, thermal power stations, and other industries and domestic use.

In the case of coal for the iron and steel industry, consideration has been given to the current target of achieving a level of steel production approximately equivalent to that of consumption and a foreign trade balance in equilibrium, and to the fact that the possibilities of expanding production will be limited until the fourth iron and steel plan comes into effect. Moreover, the iron and steel industry is expected to be virtually the only consumer of coking coal.

Demand for coal intended for thermal power stations is expected to be equivalent to the amount that can be produced in a reasonably economic manner.

In the demand forecast for coal intended for other purposes, account has been taken of consumption patterns during recent years and the fact that the demand for anthracite pellets for domestic use, heating and certain industries may be expected to remain stationary or even increase slightly.

The following shows these estimates expressed in mill.t:

	1974	1980	1985
Coal			
Iron and steel industry	6,000	9,000	12,000
Thermal electric power stations	3,310	3,196	2,796
Other	690	700	700
Total	10,000	12,896	15,496
Anthracite			
Thermal electric power stations	1,725	3,622	3,622
Other	925	1,200	1,200
Total	2,650	4,822	4,822
Brown lignite			
Thermal electric power stations	400	14,225	14,225
Black lignite			
Thermal electric power stations	2,000	8,400	8,400
Other	200	200	200

Total	2,200	8,600	8,600
OVER-ALL TOTAL	15,250	40,543	43,143

(c) Production Targets, Import Requirements and Export Policies

As a result of the expansion of the iron and steel industry and the decline in domestic coal production, coking coal import requirements increased significantly during the period 1959-1973. Virtually all coal imports - 3.1 mill.t of coking coal and 579,600 t of coal and anthracite - were used for the manufacture of coke, which was also imported.

Spanish exports were low, except in 1969 and 1970, when total exports amounted to approximately 720,700 t.

As regards production targets, on the basis of an analysis of trends in the various cost components and the output of the enterprises concerned, and of a number of alternative production forecasts, it is estimated that the production levels set forth in the following table, expressed in thousands of tons, are attainable under conditions competitive with alternative or imported sources of energy.

	1975	1980	1985
Coal	7,150	8,910	8,790
Anthracite	3,000	4,828	4,820
Black lignite	2,300	8,600	8,600
Brown lignite	1,100	14,220	14,220

Domestic supply is insufficient to meet the demand for coking coal. It has been calculated that import requirements for the years 1975, 1980 and 1985 will be 4.6, 4.1 and 6.6 mill.t respectively.

On the other hand, coal-mining activity is reaching peak levels and will account for the generation of 36,000 mill. KWh in thermal electric power stations. Approximately 10,000 mill. KWh will be derived from lignite from open mines and 5,000 mill. KWh from underground lignite having a high calorific value.

R. SWEDEN

(a) Energy Demand and Supply

In spring 1975 the Government prepared the Energy Management Bill, which was passed by the Riksdag on May 28th, 1975. The Bill states the aim of the Swedish energy policy plan which for the medium term is an energy consumption growth of an average of 2% per annum up to 1985. The long term aim from the 1990s is zero growth.

The energy conservation programme presented in the Bill consists of loans and subsidies for energy saving investments in the domestic and industrial sectors, the creation of local energy plans, educational measures, general information to the public as well as energy saving campaigns and increased energy taxes.

(b) Coal Demand and Supply

In Sweden coal is used mainly in the industrial sector, especially within the iron and steel industry. Coal can also be used in a few power stations, but due to the price uncompetitiveness of coal, coal has lately not been used for power generation. There is no production of coal in Sweden. The table below shows imports and consumption of coal by quality.

TABLE I.19 Coal Imports and Consumption (000.t)

Year	Imports			Consumption		
	Coal	Coke	Total	Coal	Coke	Total a/
1972	996	1,208	2,204	560 b/	1,840	2,400
1973	1,037	1,495	2,532	430 c/	2,083	2,513
1974	1,514	1,599	3,113	411 c/	2,391	2,802
1985	2,000	1,975	3,975	500	3,100	3,600
1990	2,000	2,375	4,375	500	3,500	4,000

a/ About 4 per cent is used for heating; the rest is for industrial use.

b/ Of which 140 was used by the Gasworks of Stockholm.

c/ Includes a small coal-based thermal power station (Hässelbyverket).

S. SWITZERLAND

(a) Definition of the Role of Coal in the Total Energy Context

It is at present difficult, if not impossible, to make a positive assertion concerning the role coal will play in the Swiss energy balance for the years 1985 and 1990. The reasons for this are given below.

Switzerland has no domestic coal production that would call for a specific policy on the part of the public sector (electric power stations) or private users.

However, future consumption of solid fuels will probably be influenced by the conclusions and recommendations of the committee responsible for preparing a comprehensive approach to energy.

It is not yet possible to say whether and to what extent Switzerland's participation in the international energy programme will affect the volume and structure of Swiss energy consumption. Coal plays an essential role as a substitute product for imported crude oil. However, its economic starting point is such that a rapid and appreciable increase in demand is not to be expected. This is attributable mainly to transport and environmental problems connected with the use of solid fuels, and to the fact that consumers are currently turning to liquid and gaseous fuels. Hence, it is only in the future and after transformation that coal might acquire importance as substitute energy.

Moreover, the trend in the price of coal on the one hand and the coal/fuel oil price ratio on the other will have a very great influence on future consumption (1985 and 1990) of solid fuels. In particular, industrial firms which are in a position to use either coal or fuel oil will continue to choose their source of energy by reference to economic considerations.

(b) Demand Forecasts, by Major Sectors and Qualities: Greater and New Markets for Coal

According to statistics published by the Federal Energy Economy Office, coal still accounted for 1.9 per cent of total energy consumption in 1974. The fact that this proportion has been steadily declining for some years, and that the energy crisis has not broken this trend confirms the minor importance of coal at present. Of the coal consumed in 1974, 60 per cent was used for domestic and handicraft purposes and 40 per cent in industry.

As to demand by major sectors and qualities and the expansion of existing, and creation of, new markets, it is very difficult - for the reasons already given under (a) - to establish forecasts for the next 10 to 15 years. No pattern is emerging from which it could be inferred that coal consumption is more likely to follow one trend rather than another during the period under consideration. From a realistic point of view, it may be expected that industry and households will still consume a certain amount of solid fuels in 1985, but the percentage accounted for by the domestic sector should decline. This trend might be influenced decisively by increased supplies of natural gas, which has certain advantages in domestic use, such as less pollution of the environment, comfort, cheap upkeep of heating equipment and space savings due to the elimination of storage problems.

Should the current situation change for any reason, the coal trade considers that coal would be suitable for the following uses, some of which are new:

- the production of electricity in electric power stations;
- the production of heat in district heating power stations;
- the production of electricity and heat in coal-fired thermal power stations;
- increased use in industry in order to reduce fuel oil consumption.

The first three uses, in installations operating on a dual-purpose basis, would also permit the use of natural gas and/or fuel oil.

These possible uses of coal are based on the assumption that the present situation will change, so that it is impossible to quantify future needs or to foresee what the future relationship between coal and other energy products will be.

(c) Production Targets, Import Requirements and Export Policies

As has already been indicated, Switzerland has no coal deposits, so that it is not faced with questions concerning production and export. As to import requirements, it is reasonable to think that Switzerland will probably turn to a greater extent than in the past to certain traditional suppliers, whereas the importance of other exporters should decline unless their coal sector is given a new impetus.

T. TURKEY

(a) Definition of the Role of Coal in the Total Energy Context

Coal (anthracite and lignite) occupies an important part in primary energy use. However, while the percentage of coal in the primary energy in 1960 was 26 per cent, it has been reduced to 20 per cent in 1973. On the other hand, the percentage of anthracite, being 19 per cent in 1960, diminished to 11 per cent in 1973, but the percentage of lignite, being 7 per cent in 1960, augmented to 9 per cent in 1973.

(b) Demand Forecasts (in mill.t)

	Demand for anthracite		
	1980	1985	1987
Electricity generation	0.535	0.535	0.535
Iron and Steel Industry	6.600	7.600	7.600
Other Industries and Heating	1.892	1.892	1.892
Total	9.027	10.027	10.027
	Demand for lignite		
Public sector	31.296	45.861	56.444
Public and private sectors	41.893	65.208	74.391

(c) Production Targets

For anthracite: the production of anthracite is expected to be 5,280,000 t in 1975 and 5,850,000 in 1980. According to known resources, optimal production for 1985 and 1987 will be also 5,850,000 t. Excess demand is expected to be 741,000 t in 1975; 3,177,000 in 1980 and 4,177,000 t in 1985. The difference will be met either through imports or by domestic substition which is under study.

For lignite: the production of lignite is expected to be 11,491,000 t in 1975, 39,830,000 t in 1980, 55,091,000 in 1985 and 64,053,000 t in 1987. Excessive demand is expected to be 2,063,000 in 1980, 10,117,000 t in 1985 and 10,338,000 t in 1987. The difference will be met either through import or domestic substitution means.

U. UNION OF SOVIET SOCIALIST REPUBLICS

The scientific institutions and planning and economic bodies of the USSR are at present preparing the five-year (1976-1980) and long-term (to 1990) economic development plans of the USSR 1/. At the time of preparation of this Note, outlines of the socio-economic concepts of development of the USSR economy during those periods had been drawn up, indicating the principal social, economic and scientific and technical problems that would be encountered in each period and the main ways and means to be used in their solution. As regards development strategy a great deal of attention is being given in these plans to the strengthening of economic, scientific and technical co-operation between the USSR and foreign countries on a mutually advantageous basis and to the implementation of a comprehensive programme of socialist economic integration.

1/ Taken into account in Table IV.2

At this stage in the formulation of long-term and five-year plans in the sphere of material production, work is concentrated on determining agreed development trends in large intra-sectoral complexes (fuel and energy, agricultural and industrial, building materials, tool making, etc.), providing for a rise in the technical level of production and quality of output, the establishment of advanced machinery and equipment systems, new, highly efficient types of materials, better use of raw materials and energy resources, as well as aggregated quantitative growth indices for individual complexes and economic sectors. Preliminary estimates have shown that it will be possible to satisfy domestic requirements as regards basic forms of energy, raw materials and goods by developing appropriate national sectors and types of production during these plan periods. At the same time it will be possible to expand the volume of supplies of various kinds to socialist countries, some developing countries and industrially developed countries of the West.

The actual volume of supplies of raw materials and fuel and energy resources furnished by the USSR to other countries during the period up to 1980 is, in general, fixed in agreements already concluded or in preparation. The volume of such supplies during the later period will be specified in long-term agreements already concluded or in preparation, and in decisions taken by the Soviet Union on a number of difficult problems connected with the development of the fuel and energy system as well as with increased output and improved use of raw materials.

In the first place, the problem is to create a large industrial potential for the fuel and energy system in the northern and eastern regions of the country where the Soviet Union's main mineral deposits, accessible to exploitation in the immediate future, are concentrated. This will require substantial capital outlay in the mining and related sectors, as well as on the transport of raw materials and energy to the western regions where the main fuel and energy consumers are at present concentrated. Exports of these resources will take place mainly from these regions. A ramified infrastructure will also have to be established in the eastern regions to assure the development of a fuel and energy system.

A number of major scientific and technical, economic and other problems will have to be solved in connexion with the planned economic growth and balanced development of the fuel and energy system during the main period. Intensified scientific research will have to be directed to the discovery of new sources of energy, the creation of new and efficient forms of energy and the improvement of production technology so as to make more rational use of fuel and energy and raw material resources. Over the long term, recent trends and the scale of structural changes in the fuel and energy balance will have to be modified. Because of the relatively limited reserves of oil and gas, more attention will have to be given to expanding the coal industry. The construction of a series of nuclear power stations will be accelerated and their efficiency will be considerably improved by using the results of Soviet scientific research.

All measures directed at increasing the volume of output and production of basic products and energy must obviously be reinforced by measures to improve the use of these resources and achieve economies in their consumption. Therefore, in working out draft plans, much attention is given to perfecting the technology and structure of production and to developing advanced equipment designs and new chemicals. Significant changes in structure and output are planned, as well as a considerable increase in capital investment for improving the technical level of production .

V. UNITED KINGDOM

(a) Definition of the Role of Coal in the Total Energy Context

In the United Kingdom the coal industry has a major role to play in the long-term strategy for meeting the country's energy needs. Despite the decline in production in recent years coal still provides a substantial proportion of the energy needs, i.e. 37 per cent in 1974/1975.

In recognition of this and the changed energy situation following the crisis of late 1973 the Government in 1974 carried out a detailed examination of the coal industry together with the National Coal Board and the mining unions. The object was to advise on the contribution coal could best make to the country's energy needs and the steps needed to secure that contribution. In two reports (an Interim Report 1/ published in June 1974 and a Final Report 2/ in November 1974), the examination concluded that the new energy situation did indeed provide greater market opportunities for the United Kingdom coal industry but subject to the continuity of assured supplies at costs that were competitive in the longer term with other sources of energy.

In particular the examination confirmed the National Coal Board's plan to arrest the previous decline in production by maintaining, and possibly expanding, current production in the period to 1985 by:

- investing in 42 mill. long t of new deep-mined capacity by extending the life of existing pits; increasing output by major developments at certain pits and opening new pits; and
- increasing opencast coal production by 50 per cent from 10 to 15 mill. long t a year.

The Government fully endorses these conclusions and has accepted the National Coal Board's Plan for Coal. This involves additional investment of some £600m at 1974 prices in the United Kingdom coal industry over the next decade bringing the total investment over that period (including the normal capital requirements of the industry) to some £1,400m. In order to ensure steady and sustained progress in the implementation of these plans the Government has given an assurance that short-term fluctuations in the price and availability of competing fuels will not be allowed to interfere. This undertaking however recognizes that in the long term coal must remain competitive in order to make the fullest possible use of the opportunities now open to it.

The fact that the United Kingdom has substantial resources of gas and oil as well as coal is not in any way regarded as a threat to these opportunities for the coal industry. The reserves of coal are larger than those of other fuels and there is no question of the United Kingdom being so rich in indigenous resources over-all that the value of coal reserves will be under-

---

1/ Department of Energy, Coal Industry Examination - Interim Report, London, June 1974.

2/ See 1/, Final Report, London, November 1974.

estimated. Coal's position in the United Kingdom over-all energy strategy is thus both substantial and assured.

(b) Demand Forecasts and New Markets for Coal

Forecasts of demand over 10 or 15 years ahead are subject to considerable uncertainty. The Department of Energy has carried out, and is continuing extensive studies over the whole energy field in order to minimize such uncertainty as far as possible. This work covers the many alternative assumptions that have to be taken into consideration. The resulting calculations have demonstrated the very wide range of possible forecasts that may result.

While it is not therefore possible to give precise forecasts of coal's contribution to United Kingdom energy requirements it appears at present that overall demand in 1985 will be of the order of 135 mill. long t.

Research into coal conversion processes is being energetically pursued in the United Kingdom (which is co-operating fully in a number of international research projects in this field). However it is unlikely that there will be a significant demand for coal for this purpose within the next decade. Nor is such a demand likely to be very significant by 1990. Thereafter however there could be a growing demand for coal and for coal conversion processes, but this will depend on the total world availability of oil and the oil/coal price relationship.

(c) Production Targets, Import Requirements and Export Policies

The total United Kingdom coal production in the year ending 31 March 1975 was 125 mill. long t (115 mill. long t deep-mined; 9 mill. long t opencast and 1 mill. long t from small licensed mines). This was somewhat lower than expected because of the after effect of the 1974 strike in the industry. The aim over the next decade is at least to maintain deep-mined production at around 120 mill. long t a year. This will be done by preventing the inevitable decline that would otherwise occur by investing in new and improved capacity to counteract the normal closures resulting from exhaustion of seams and other technical difficulties at existing collieries. This, plus the planned increase in opencast production to 15 mill. long t a year, should enable the expected 1985 demand to be covered, particularly if optimum use can be made of existing capacity and all new projects are brought in on time.

United Kingdom imports of coal in recent years have been very small (between 1 and 3 mill. long t a year) except in those years (1972/1973 and 1974/1975) when home supplies were affected by industrial disputes. Imports are currently a little higher than normal because of short-term commitments entered into at the time of the last dispute in 1974. They are expected to revert to more normal levels shortly. Imports normally consist mainly of certain qualities of coking coal (to supplement by blending with home production for the United Kingdom iron and steel industry) and some steam coal.

The United Kingdom has traditional customers for certain qualities of coal which have been regularly exported. Exports have ranged between 2 and 3 mill. long t a year over recent years. The aim is to maintain these and progressively expand them as the United Kingdom's over-all coal supply situation improves.

W. UNITED STATES OF AMERICA

The medium-term energy policies shift the focus to the development of new sources of energy and more sustainable conservation efforts. Simply stated, the goal is to eliminate the threat to economic and national security resulting from a disruption of foreign supplies of petroleum. Since petroleum accounts for approximately one half (46 per cent in 1973) of the total United States energy needs this risk is obviously a serious one. If no actions are taken the domestic consumption of oil would rise to 23.9 million barrels per day by 1985 and imports would represent 53 per cent of that total. If Congress will pass the necessary legislation the recommended programme can achieve the general goal by 1985. The specific actions to achieve the President's medium-term goal are summarized in Table I.20.

The beginning step in the medium-term programme is to increase the productivity of existing petroleum resources. The elimination of price controls over domestic crude oil provides strong incentives to produce more oil from existing wells through improved recovery procedures. The pace of exploration and drilling in traditional areas has also accelerated in recent months. However, the major share of the increased output will have to come from new wells drilled in the outer continental shelf and Alaska. Extensive development has already occurred off the coasts of Texas and Louisiana but discussions with other States for new projects are now being held. The Federal Government will be leasing new offshore drilling sites that should provide 1.5 million barrels per day of oil by 1985.

TABLE I.20 Effects of Mid-Term Programme - 1985
(million barrels per day)

Demand with no new actions		23.9
Imports with no new actions		12.7
Savings achieved by new actions -		
Other continental shelf leasing	1.5	
NPR-4 reserve development	2.0	
Coal conversion	0.4	
Synthetic fuels	0.3	
Auto efficiency standards	1.0	
Proposed taxes and import fees	2.1	
Appliance efficiency goals	0.1	
Insulation tax credit	0.3	
Thermal standards	0.3	
Total savings		8.0
Remaining imports		4.7
Emergency storage		3.0
Standby allocation power		1.7
Net import vulnerability		0

Legislation is also before Congress to begin development of the Naval Petroleum Reserve (NPR-4) in Alaska. This is a very large oil field that should provide 2 to 3 million barrels of oil per day and comparable quantities of natural gas by 1985. Development of the Alaskan pipeline is finally moving ahead after a series of unfortunate legislative delays. Approximately 20 per cent of the output will be allocated for strategic military and economic reserves.

Major attention will be given to developing the massive $1\frac{1}{2}$ trillion t of coal reserves which are our most abundant energy resource. The output of coal has been declining for many years as home owners and industrial users have shifted to low-cost oil and natural gas sources of energy. Total coal output is expected to double by 1985 to meet growing energy needs. Conversion back to coal burning furnaces is expected to alleviate part of the oil import pressures. Congress has been asked to temporarily suspend certain air quality standards until low sulphur coal becomes available and the technology for utility stack gas scrubbers can be developed. Legislation is currently before Congress to authorize surface mining and the output of deep mines should increase given the current energy requirements. The Federal Government is accelerating the leasing of Federal land holdings containing coal reserves.

TABLE I.21 Growth of Energy Consumption 1973-1990
(assuming an average price of $7 per barrel of oil)

10^6 tce

	1973	1977		1980		1985		1990	
		WOC	WC	WOC	WC	WOC	WC	WOC	WC
Petroleum	1,230	1,340	1,260	1,470	1,360	1,700	1,510	1,960	1,740
Natural gas	807	765	750	810	775	845	800	885	835
Coal	475	600	590	640	600	705	590	775	630
Hydro-geothermal	34	40	40	47	47	56	56	66	66
Nuclear	11	33	33	56	56	147	147	254	254
Total	2,557	2,780	2,670	3,020	2,840	3,450	3,100	3,940	3,520
Petroleum imports	220	...	...	340	285	440	345	560	460

WOC = without conservation.
WC = with conservation

TABLE I.22 Growth of Energy Consumption 1973-1990
(assuming an average price of $11 per barrel of oil)

10^6 tce

	1973	1977		1980		1985		1990	
		WOC	WC	WOC	WC	WOC	WC	WOC	WC
Petroleum	1,230	1,210	1,150	1,240	1,150	1,350	1,185	1,460	1,290
Natural gas	807	770	770	820	785	880	840	940	890
Coal	475	590	580	690	665	810	700	950	820
Hydro-geothermal	34	40	40	47	47	56	56	66	66
Nuclear	11	33	33	56	56	147	147	254	254
Total	2,557	2,640	2,570	2,850	2,700	3,240	2,930	3,670	3,320
Petroleum imports	220	...	...	163	116	116	42	138	64

Source: Federal Energy Administration, Project Independence; Washington 1970; original figures have been converted at 1 barrel a day = 75 tce a year; 1 kWh = 860 kcal.

With regard to the potential and prerequisites of coal exports from the United States and to the expected increase of western European coal imports by 35-40 mill.t by 1985 (see Part II of the present study, table II.1), the United States Government made the following statement at the thirteenth session of the Working Party on Coal Trade (COAL/WP1/28, para. 25).

> "The bulk of the additional 35 mill.t of coal imports estimated for 1985 could be supplied by the United States. As a matter of fact, published estimates of individual exporters and collective estimates of future export levels indicate that the United States could accommodate all of the additional demand. Although such an increase would doubtless be spread over as many suppliers as could be deemed reasonable to diversify sources of supply, the share falling to the United States suppliers would necessarily have to be arranged on long-term commitments for which new productive capacity would have to be developed. From the standpoint of the adequacy of resources and infrastructure, however, there appear to be no reasonable grounds to doubt the capability. The actual planning and execution of a large increase in coal exports to Europe would depend upon the collective economic aspects of the new export coal development. These considerations are also predicated upon the resolution of some of the more pressing immediate problems facing the coal industry today which relate to national and local environmental protection objectives and other national policies which would facilitate new coal mine development."

X. YUGOSLAVIA

The fast economic development of Yugoslavia in the years after the Second

World War, especially the increase in industrial production, conditioned a considerably rapid development of energy consumption requiring at the same time the production increase of all kinds of primary energy and coal production. In the past the largest portion of the total energy consumption in Yugoslavia had been met by coal. Compared to 1960 the consumption of all kinds of primary energy increased 2.2 times compared to 1972 and 3 times to 1955.

TABLE I.23 Structure of Primary Energy Consumption (in percentage)

Year	Total	Solid fuel	Liquid fuel	Natural gas	Hydro-energy
1960	100	80.8	13.8	0.5	4.9
1965	100	71.1	21.4	2.1	5.5
1970	100	55.0	36.1	2.5	6.4
1972	100	49.5	41.0	2.9	6.6
1973	100	49.2	41.8	3.4	5.6

Energy consumption in Yugoslavia reached approximately 33 mill. tce in 1973, while it was 15.05 mill. tce in 1960 and 21.36 mill. tce in 1965. Along with consumption the production realized in 1946 was only 6.65 mill. tce. Six years later, in 1952, the production increased 2 times and 19 years later, in 1964, 4.5 times. With coal production of 29.9 mill. t in 1965, a period of stagnation started and production declined up to 1970 when the coal production was 30.9 mill. t, and in 1973 it amounted to 32.4 mill. t.

Two periods were characteristic for energy consumption in Yugoslavia, the period up to 1965 and after 1965. The following table gives the actual average growth rates by single types of primary energy.

TABLE I.24 Realized Average Growth Rates (in percentage)

	1955-1960	1960-1965	1966-1970	1971-1973
Coal	6.2	6.8	1.6	3.7
Crude oil	11.6	16.9	16.1	9.1
Natural gas	13.7	40.0	21.3	7.7
Hydro-energy	11.8	8.1	9.8	1.5
Total	8.0	8.1	4.8	5.8

In spite of a production decrease in the period 1966-1970 both in the absolute amount and in its relative participation in the total demand, the coal industry in Yugoslavia played an important role in the national economy during the past period. It supplied even up to 80 per cent of the total Yugoslav energy consumption. In 1973 (though the importance of coal had relatively declined), the participation of coal in satisfying the energy demand had been about 50 per cent and is still the most important as regards energy raw materials.

From the data given above it can be seen that the total increase in energy consumption with a smaller share of coal was made up with the high consumption increase of crude oil derivates, resulting in a great increase in fuel imports, mostly crude oil, because the domestic crude oil production covered only about 35 per cent of the total demand in 1973.

Based on data concerning energy consumption in 1973, it is estimated that the percentage of dependence of the country on imports was about 31.6 per cent. It is to be emphasized that this dependence had been permanently growing because in 1965 it was 17.2 per cent. Nevertheless this percentage of energetic dependence of the country is considerably lower than that of developed countries of western Europe (this percentage in OECD member countries was 58.8 in 1970). It shows that the consumers were orientated towards a primary energy consumption which permitted a higher productivity, better quality of products and at the same time simplified the handling of technical processes and a more effective production. Therefore, due to these advantages and also to the availability of crude oil and natural gas as well as electro energy on the market a significant portion of the industry was using the new technology. The protection of the environment also required the change-over of certain consumers to an energy source without or with a small amount of poisonous gases, ash and other harmful compounds which pollute the atmosphere.

Yugoslavia owns relatively significant energy reserves, of which coal covers the biggest portion:

TABLE I.25 Structure of Energy Reserves

Coal	84.0
Crude oil	1.4
Natural gas	1.1
Hydro-energy	11.5
Nuclear fuel	2.0
Total	100.0

As regards the characteristics of the energy resources, it is assessed that approximately 50 per cent of available hydro potentials (which amount to 2/3 of the total energy potential) is in use or under construction, the remaining less economical portion of hydro-energy sources will be built in the years to come with the co-operation of other water users.

According to the present estimation there are sound prospects that coal and hydro-energy will make the fundamental energy base for electric energy production, while other resources will be gradually included, depending on the economic policy of the country, the result of further studies, geological and other surveys.

Crude oil and gas make up for 2.5 per cent of the reserves. Regarding the stage of exploration it can be expected that the discovery of new crude oil and gas reserves in the Panonian basin, and also the intensive exploration in the future will give more reliable data on possible reserves of crude oil and gas in the Adriatic region.

The total reserves of all kinds of coal in the country are estimated at approximately 48,420 x 10^{12} kcal. Lignite has the greatest share of the total coal reserves in Yugoslavia, both in physical volume (90.2 per cent) and in calorific balance (82.2 per cent). Brown coal provides 9.3 per cent in physical volume and 16.4 per cent in calorific value, while the reserves of bituminous coal are small, making only 0.5 per cent.

Regarding either international or domestic classification, the coal does not belong to the good quality groups. The bituminous type of coal is characterized by high sulphur contents and some also by high ash. Brown coals also have a higher sulphur (1-5 per cent) and ash (10-25 per cent) content, in fine commercial grades of coal even up to 40 per cent. The lignites contain much moisture (often up to 56 per cent) and ash (8-17 per cent). But a very favourable situation is that about 92 per cent of total exploitable reserves regarding both volume and calorific value are concentrated in around ten coal basins, where 65 per cent of total reserves can be mined by open-cast operation; concerning lignite it is 80 per cent.

Coal reserves in the country are very important, and very promising, from which an economically favourable coal production of about 180-200 mill. t per year can be organized. From the technical, geological and economical points of view coal availabilities have up to now exceeded, and will probably in the near future exceed the actual needs. The structure and allocation of the deposits permits the direction and future development of coal production towards a greater concentration and application of modern mechanization as well as the introduction of automation in order to obtain the lowest production costs.

On the basis of approximate parameters resulting from the study of long-term development of Yugoslavia by authorities, institutions and various responsible bodies, it can be concluded that the energy consumption will considerably go up. The long-term forecasts of energy development in Yugoslavia are based on the following:

- well balanced energy development as a whole, and its successful use in the economy;
- maximum use and development of the domestic energy base; long-term and reliable production and adequate quality of all kinds of energy at the most favourable prices for consumers; at the same time taking into consideration economic, strategic and other aspects;
- optimal use of existing capacities and the wider and long-term co-operation of the producers and consumers on the basis of self-managing agreements;
- long-term expansion of the productive forces which will stimulate rational solutions; provision of large energetic systems and the construction of complex energetics capacities based on modern scientific-technical achievements and the application of modern techniques; modern technologies and a better organization of production will enable better managing results and such energy prices which will permit a successful participation in foreign trade and other forms of international economic collaboration.

According to these forecasts the energy consumption will multiply. The structure of primary energy consumption is shown in the table.

TABLE I.26 Structure of Primary Energy Consumption up to 1990

Year	Total	Solid fuel	Liquid fuel	Natural gas	Hydro-energy	Nuclear energy
1975	100	30.0	39.0	11.0	20.0	-
1980	100	27.0	40.0	14.0	17.0	2.0
1985	100	28.0	41.0	12.0	16.0	3.0
1990	100	29.0	40.0	12.0	15.0	3.0

The energy consumption will reach 68.20 mill. tce in 1980, and approximately 124.30 mill. tce in 1990, i.e. will double in the next 15 years. In addition, electricity consumption will reach approximately 145 milliard kWh, coal consumption about 130 mill. t, crude oil 36 mill. t, natural gas 12 milliard Nm^3 and nuclear fuel about 200 tons (U_3O_8).

The forecasted energy consumption will be satisfied by exploration of domestic energy resources and also by the development of domestic production of secondary forms of energy on the basis of the transformation of domestic and imported (crude oil and gas) primary energy sources. Electricity consumption will be provided by domestic production, coal consumption too (except coking coal), while the crude oil demand will be satisfied by domestic production (about 8 mill. t, i.e. 35 per cent) and the increased import of about 28 mill. t in 1990 compared with 7.2 mill. t in 1975, and also by gas imports (at present it is not imported) in considerable amounts (about 600 mill. t Nm^3 in 1990).

The data foreseen might be modified through further discussions possibly in favour of domestic resources mainly of coal which would reach a consumption level per capita of about 4,900 kg of coal equivalent in 1990 compared to 1,540 kg of coal equivalent in 1973. According to present criteria based on specific energy consumption, the country will be put in the group of relatively developed countries as soon as 1985.

From data concerning the prospective consumption of all kinds of energy it can be seen that coal will take a significant place, although its participation in the total consumption will be lower than now. On the basis of coal reserves it is possible to organize, as we have already mentioned, a considerably higher production. However, coal production, i.e. the coal mines capacities, will be developed in accordance with demand.

The policy of energy production, which is based on consumption and on analyses carried out by many institutes and other organizations, shows that coal consumption in the domain of industry will be mostly represented by thermal power plants, heating plants and boiler rooms close to the coal basin. There the coal price can be competitive with other types of fuel. In the transportation sector, the coal will be completely eliminated. In the general and household consumption along with the growth of living standard, coal consumption will decrease in towns, while in the relatively small settlements, where the coal price is competitive to other fuels, an increase in coal consumption can be expected.

The total forecasted coal consumption (except coking coal), as we have already emphasized, should be provided by indigenous production. The production concentration in coal mines was intensified after 1966 by building bigger capacities and closing down smaller mines. More significant steps in production concentration have been taken by the big mining basins. Up to now 45 coal mines have been shut down, partly due to the regional concentration and partly due to unprofitable operations or exhausted reserves. The supposition is that coal production will be limited to big basins. Coal will be consumed on the spot by thermal power plants. The big prospective mines (lignite mines with open-cast operation and underground mines with good quality brown coal), have undertaken a significant investment effort in constructing the most modern and economical capacities, though the optimal capacities have not been reached yet. They guarantee that in the years to come a satisfactory effect will be realized permitting more successful operation and assuring greater funds for the expansion of the productive forces and the higher living standard of employees.

Y. NON-ECE COUNTRIES: AUSTRALIA 1/

(a) Forecast Demand for Energy

Australia's demand for primary energy increased at an average rate of 5.7 per cent per annum in the period between 1960-1961 and 1973-1974. It is expected that the rate of increase might decline marginally (to perhaps 5.5 per cent per annum) between 1973-1974 and 1984-1985. The tables below show recent estimates of total primary energy consumption in Australia to the year 1984-1985, in both energy units and appropriate physical units. (The forecasts were prepared in the Australian Department of National Resources. They are, however, not to be considered as an expression of the Australian Government's view as to the likely outcome, or the path the Government would necessarily wish the fuel economy to follow).

TABLE I.27 Australia's Demand for Primary Fuel: In Energy Terms

(Thousand terajoules)

Fuel	1970-1971		1975-1976		1980-1981		1984-1985	
	Qty.	%	Qty.	%	Qty.	%	Qty.	%
Petroleum	1,066	49.8	1,348	48.1	1,721	45.7	1,832	40.5
Natural gas	74	3.5	217	7.7	417	11.1	694	15.3
Black coal	654	30.5	818	29.2	1,143	30.0	1,411	31.2
Brown coal	228	10.6	293	10.5	357	9.5	450	9.9
Hydro-electricity	42	2.0	46	1.6	50	1.4	58	1.3
Wood	39	1.8	33	1.2	27	0.7	23	0.5
Bagasse	40	1.8	49	1.7	51	1.5	59	1.3
Total	2,144	100.0	2,803	100.0	3,769	100.0	4,528	100.0

1/ Australia participated in the work of the Working Party on Coal Trade.

TABLE I.28 Australia's Demand for Primary Fuel: in Natural Units

Fuel	Unit	1970-71	1975-76	1980-81	1984-85
Petroleum	Thous. bbls.	206,348	259,141	327,232	352,311
Natural gas	Million cu. metres	1,913	5,572	10,725	17,850
Black coal	Thous. ton	24,956	31,382	44,248	54,887
Brown coal	Thous. ton	23,350	29,616	36,012	45,461
Hydro-electricity	Million kWh	11,838	12,722	14,028	16,222
Wood	Thous. ton	2,413	2,037	1,648	1,438
Bagasse	Thous. ton	5,247	6,328	6,601	7,682

The forecasts indicate that a degree of energy substitution will occur over the next decade or so, in that the proportion of coal and natural gas consumed in Australia will increase at the expense of petroleum.

Of the total primary energy demand in 1973-1974, the manufacturing sector accounted for 33 per cent, electric utilities 28 per cent, the transport sector 27 per cent, and the domestic/commercial sector for 6 per cent. Coal supplied 41 per cent of the total demand in the manufacturing sector, 85 per cent for electricity generation, and 20 per cent in the domestic/commercial sector.

The future demand for coal for electricity generation is expected to increase as a result of increases in fuel oil prices. Additional black coal-fired power stations are being built to cater for internal demand. It has been projected that the use of black coal for electricity generation may increase from 15 mill.t in 1973-1974 to approximately 25 mill. t in 1985.

The other major source of demand for coal is the steel industry, and the anticipated long-term growth on this sector should ensure an expanding local market for Australian coking coal. It is estimated that black coal consumption by the steel industry could reach 16 mill. t by the mid-1980s and 30 mill. t by the year 2000, compared with the 1973-1974 usage of 9 mill. t.

One uncertain factor in the future internal demand for coal is the degree of development of coal gasification and liquefaction processes. An illustration of the importance of this technology to future demand for coal is provided by the fact that about 105 mill. t of coal would be required annually to satisfy Australia's present consumption of oil of approximately 620,000 barrels per day. It is, however, not expected that there would be significant developments in this area until perhaps the mid-1980s.

In the past, coal has traditionally been Australia's most significant energy export. In 1974/1975 coal became one of Australia's largest single export earners, exports rose in that year to the record level of 32.4 mill. t, compared with shipments of 28.4 mill. t in 1973-1974 and 25.8 mill. t in 1972-1973. To meet Japanese requirements along, exports are expected to increase by perhaps a further 20 mill. t per annum by 1980.

(b) Energy Reserves

The table below summarizes the most recent assessment of Australia's proved and probable energy reserves.

TABLE I.29 Australia's Energy Reserves, 1975

Fuel	Unit	Proved reserves (a)	Probable reserves (b)	Total, proved and probable reserves	Annual production
Black coal	mill. t	34,739	192,878	227,617	64.1 (c)
Brown coal	" "	67,100	55,100	122,200	27.4
Total coal	" "	101,839	247,978	349,817	91.5
Crude oil	" "	714	n.a.	n.a.	19.3
Natural-gas Liquids	" "	171	n.a.	n.a.	2.5
Natural gas	billion cu. metres	1,478	n.a.	n.a.	5.1
Uranium	ton U_3O_8	286,540	95,300	381,840	-
Hydro-generation	Installed capacity 5,535 MW		Under construction 1,147 MW		Annual generation 13 Billion kWh

(a) Estimated quantity of fuels in place of known deposits, based on specific sample data, measurements of the deposits and detailed knowledge of the quality or grade of the deposits.

(b) Indicated or inferred resources additional to known deposits, including those which could exist in unexplored extensions of known deposits or in undiscovered deposits in known fuel bearing areas, and those inferred through knowledge of geological conditions and the results of geological and exploratory work.

(c) Saleable production.

(c) Coal Reserves

Australia's black coal reserves may be classified into two categories, the first representing reserves which are mineable under current economic conditions and using current mining technology from coal seams which are well defined, and the second representing reserves in coal-bearing areas where drilling has been inadequate and for which data is sparse. At present, estimated reserves for the first category are approximately 30,000 mill. t and reserves for the second category are approximately 200,000 mill. t. Of the reserves in the second category, very little is suitable for open-cut mining and total open-cut reserves of black coal are currently estimated at less than 4,000 mill. t.

In New South Wales, coal resources have been best defined in the Sydney Basin, although the need for additional information, even in this area, is illustrated by the small proportion (16%) of total reserves classified as first category. Concerning the other basins of New South Wales little is known, although they are generally regarded as being inferior to the Sydney Basin coal in terms of coal-seam thickness and ash content.

The coal reserves of South Australia and Western Australia are of sub-bituminous quality. Much of the coal in Queensland is high quality coking coal by international standards, although , as in other parts of Australia, more drilling will be required to properly define the reserves in terms of both quality and quantity. The Jurassic coals of the Clarence-Moreton Basin in Queensland are strongly per-hydrous, a factor which is favourable for any future liquefaction and gasification.

The heat content of Eastern Australian black coals ranges from 23 to 28 x 10^9 joules per t and is quite high by world standards, while the black coals in South Australia and Western Australia have typical heat contents of 14 x 10^9 and 20 x 10^9 joules per t respectively.

Australia's brown coal reserves are contained wholly within Victoria, and a high proportion of these are in the LaTrobe Valley. Of 122,000 mill. t of total reserves, about 12,000 mill. t are recoverable with present mining techniques. This coal is of inferior value for heating (10 x 10^9 joules per t) and is used almost entirely for electric power generation.

Australia's reserves of both brown and black coal are considered to be more than ample to meet domestic needs for coal into the foreseeable future, and to provide a considerable margin for export.

(d) Export Potential

It is noted that the Secretariat has estimated an increased coal import requirement for Western Europe of 35 mill. tce. by 1985. Australia would be capable of supplying a major share of the requirement. Of particular significance is the fact that currently about half of Australia's production of saleable coal is destined for export, and the proportion is increasing. In view of the large resources in relation to domestic needs, it can be taken that the principal factor in triggering the development of new mines is the demonstrated availability of long term markets.

PART II

PERSPECTIVES OF THE COAL INDUSTRY IN WESTERN EUROPE

In Part I of the present study Governments have exposed their policies with respect to the role of coal in meeting energy demand in the next decade. Part II of the study analyses these policies with a view to identifying general trends and emerging problems with respect to Western Europe as a whole.

A. REASONS FOR CHANGE

(a) A Host of Motivations

Contrary to what is commonly held the reassessment of the role of coal undertaken by Western European Governments in 1974 and 1975 had been motivated by a host of motivations among which the "oil crisis" of 1973/1974 was certainly the predominant factor, but not the only one. The other factors were the unsatisfactory progress of nuclear power in Western European electricity supply systems, the growing concern about the timely availability of breeders, doubts about a further continuation of the growth-orientation of the economic policy, and a greater commitment towards the preservation of natural resources, energy and the human environment in general. These factors might well have prompted earlier or later a renewed attention on coal as a form of energy which could possibly bridge the gap between the present energy supply pattern and any future supply pattern based primarily on fission, fusion, solar power or any other energy mix.

(b) Emerging Policy Reactions

It is striking to note the similarity of the situations in which Western European countries found themselves in the years 1973 and 1974 with respect to these problems. It seems that the increased reliance on the world oil market and on the development of one particular technology - fission - has created a degree of interdependence between these nations which was hitherto unknown. It is even more striking to note the similarity of solutions aimed at by Western European Governments in these years - a similarity which contrasts with the diversity of approaches witnessed in the 1950s and 1960s. These new and widely shared endeavours can be summed up as follows:

- finding a new equilibrium between the demand for continued economic growth and the demand for a better quality of life and for greater financial stability;
- increasing the efficiency of raw material and energy use in order to preserve these resources for future generations and to reduce the discharge of pollutants into the human environment;
- reducing the reliance on imports of oil from outside the region;
- updating earlier coal production and import targets in order to increase the security of supply and to bridge the gap until the solution of the problems presently facing

nuclear power and encountered with the extraction and shipment of North Sea gas and oil.

Two conclusions can be drawn from the outset:

- the new role of coal has been defined in the context of the energy and economic policies; it was not coal as such which Governments were looking for, but an increase in the security of supply and a means to "buy time";
- the new tasks are of a medium-term nature; the uncertainties of the future did not permit a quantitative forecast beyond 1985 particularly with regard to non-coking coal.

B. RESULTING ENERGY DEMAND GROWTH RATES

The reorientation of general economic policies and the increased commitment to a higher efficiency of resource use are likely to substantially slow down the growth rate of energy demand in Western Europe. This slowdown will be the result of two factors: a slowdown in economic growth and an increase in the efficiency of energy use per unit of gross domestic product.

(a) Reduced Rate of Economic Growth

The annual growth rate of GDP in the Western European market economies has decreased from 4.8 per cent during 1953-1960 to 4.7 per cent in 1961-1970 and 4.5 per cent in 1968-1975. The Secretariat estimates that the major factors governing GDP growth are likely to lead to a further slowdown in this growth rate. These factors are continued anti-inflationary government policies, limitations in the availability of manpower in Northern and Western Europe resulting from demographic and social factors, constraints in the availability of capital as a result of changed savings behaviour or savings capacity, and a greater emphasis on the service sector, on better environmental conditions, earlier retirement age, etc. Making allowance for one or two recessions and increased investments in the "non-productive" sphere for environmental control or other purposes, the secretariat suggests a plausible growth rate between 1976 and 1990 of:

1.8 to 2.3 per cent for Northern Europe
3.3 to 4.3 per cent for the industrialized countries of Western Europe
5.5 to 6.5 per cent for Southern Europe
3.2 to 4.2 per cent for total Western Europe

(b) Increased Efficiency of Energy Use

The deceleration of economic growth would by itself already lead to a slowdown of the rates of growth of energy demand in a similar order of magnitude since the energy/GDP elasticity is presently in Western Europe at about 1: an increase of GDP by one unit requires an additional input of energy of one unit.

However, there are three tendencies which suggest that the growth rate of energy demand will decelerate faster than that of GDP:

- a shift in the industrial structure from high energy-intensive production to less energy-intensive activities such as services;

- the determination of governments to increase the economy and efficiency of energy use;
- the growing (political and demographic) difficulty in substituting labour by energy.

A 1976 Secretariat study 1/ inquired into the first two trends mentioned. This study considers it realistic that between now and the early 1990s the economy and efficiency of energy input (including some changes in industrial structures) could be raised by as much as one-third, the theoretical maximum being a rise of 100 per cent. Since the great majority of ECE Governments announced and supported measures aiming at energy conservation, a substantial impact on the growth rate of energy demand can be anticipated.

(c) Lower Rate of Growth of Energy Demand

Under the combined effect of the two possible trends just mentioned, the growth rate of energy demand is unlikely to remain at the level of 4.7 per cent experienced during 1961-1970. Assuming an improved energy/GDP elasticity of 0.85 by 1985 as a result of increased efficiency of energy use, the growth rate of energy demand would fall to (3.2 to 4.2 x 0.85) 2.7 to 3.6 per cent.

Figures supplied by governments compare with this aggregate as follows:

France	5.2% (1961-1970); 3% (1974-1985)
Fed.Rep. of Germany	4.4% (1961-1970); 3.2% (1973-1985)
Netherlands	8% (1961-1970); 4.5% (1974-1980); 2.2% (1980-1985)
Norway	6% (1961-1970; 3.3% (1974-1985)
Spain	4.5% (1961-1970); 5.9% (1973-1985)
Sweden	6.8% (1961-1970); 2% (1973-1985)
Yugoslavia	4.1% (1961-1970); 6.2% (1980-1990)
European Economic Communities	5.6% (1961-1970); 3.8% (1973-1985)

C. ANTICIPATED RELATIVE IMPORTANCE OF COAL IN 1985

The share of coal in the total energy supply fell continually in Western Europe from 88 per cent in 1950 to 32 per cent in 1972. At the same time the share of imports in meeting energy demand rose from 20 per cent to 65 per cent. Imported oil and natural gas replaced indigenous coal as a result of greater attractiveness in terms of cost and easy handling. While this substitution was beneficial for Western Europe and greatly contributed to its growth and prosperity, it also led to a degree of dependence on foreign oil supplies whose disadvantages became drastically apparent during the "oil crisis" of 1973/1974.

The national reports reproduced in Part I describe the reassessment of the energy and of coal policy which was undertaken in Western Europe in the light of this experience. Considering also the lower-than-expected progress in the penetration of nuclear power in the Western European electricity markets, governments, apart from enhancing the development of nuclear power, proceeded

---

1/ UN, Study on measures taken, or which might be taken, to achieve increased economy and efficiency in the extraction, conversion, transport and use of energy in the ECE region, New York 1976.

to a re-examination of the competitiveness of coal (indigenous and foreign) within the overall energy context, that means with respect to indigenous oil resources (North Sea), natural gas, hydro power and new resources of energy. This examination stretched over the medium term only (1985). Whilst anticipating a long-term market for coking coal, some governments held that the long-term prospects of steam coal, in particular the prospect of its gasification and liquefaction could not yet be assessed in operational terms. Others held that the penetration of North Sea oil and nuclear power into the Western European energy market would gain such an importance by 1985 that coal could no longer cover markets in which it had not a genuine competitive advantage over nuclear power, oil and natural gas.

The redefinition of coal's medium-term role will lead to an increase in the share of coal in the total energy supply in 1985 in comparison with earlier plans for that year. This change does not, however, prevent the share of coal from falling (slowly) between now and 1985 in Western Europe as a whole. The exceptions are Finland and possibly Turkey where the share of coal is likely to increase. For Western Europe as a whole, the share of coal in 1985 will be approximately 20 per cent, compared with 26 per cent in 1974 and 11-13 per cent as foreseen before the "oil-crisis". This policy will slow down the degree of dependence on foreign oil. The European Economic Community estimates for its members that the contribution of oil imports to meeting total demand could fall from 59 per cent in 1973 to 50 per cent in 1985 (although mainly as a result of the development of nuclear power and North Sea gas and oil).

D. CHANGING COAL DEMAND PATTERNS

(a) The Total Increase

As can be seen from table II.1, the reassessment of coal's role will lead to an increase in the solid fuel consumption of approximately 11 per cent in Western Europe between 1973 and 1985. The net increase in absolute terms is about 40 per mill. tce.

(b) Deliveries to Power Stations

Most of the additional consumption is earmarked for power generation and district heating, where coal is expected to enjoy a comparatively high degree of competitiveness. The table shows that in 1985 250 mill. tce. will be used for these purposes, i.e. 55 mill. tce. more than in 1973. Power generation and district heating will account for 56 per cent of total coal consumption in 1985, as compared with 48 per cent in 1973 and 30 per cent in 1967.

(c) Iron and Steel Industry and Coke Ovens

The iron and steel sector and coke ovens will be a slightly growing market for coal. Apart from increased use of formed coke, governments do not foresee revolutionary changes in pig iron and steel making technologies until 1985; also specific coke consumption in the blast furnace is not expected to improve as long as the price relationship between coke, oil and natural gas does not change substantially to the detriment of coal. Deliveries to this sector will be approximately 115 mill. tce. in 1985, i.e. about 5 mill. tce. more than in 1973. In 1985 this sector will consume 26 per cent of the total coal supply compared with 28 per cent in 1973 and 27 per cent in 1967.

(d) Households and Others

While continuing to consume less coal all other consumers, namely households, industry, construction and others remain a substantive client for coal; in 1985 they will still account for about 18 per cent of the market for coal against 24 per cent in 1973 and 43 per cent in 1967. The loss in absolute terms is 15-20 mill. tce., which brings deliveries to this sector down to about 75-80 mill. tce.

E. STRUCTURAL CHANGES OF COAL SUPPLIES

From what sources might the additional internal demand of 40 mill. tce be covered? 81 per cent will be supplied by indigenous sources both through an increase in production and through a reduction of exports;19 per cent will be looked for in the world coal market. The strikingly high reliance on imports reflects the need mainly for coking coal and coke, but also the expectation that the improved position of coal might be a transitional one not justifying expensive investments in the domestic coal industries.

Another salient feature is the great increase in open-cast production (that is, with the exception of the United Kingdom), of brown coal and lignite. The increase in production is about 50 mill. tce. In 1985, of the total coal production 31 per cent will be from open-cast mines against 18 per cent in 1973. About 90 per cent of this production will go to power stations.

The achievement of these new tasks of substantially increased open-cast output and coal imports requires intergovernmental and inter-industry co-operation with a view to facilitating the transfer of open-cast mining technology and of optimizing the logistics of increased international coal trade (inquiry into the availability on the world market of the quantities required, shipping tonnage available, prospects for greater coal vessels, expansion of port facilities in particular with respect to the careful transhipment of coking coal, coal pipelines, standardization and simplification of trade procedures and practices, etc.).

F. NEW REGIONAL CONSUMPTION/PRODUCTION PATTERNS

In 1973 Greece, Italy, Portugal, Spain, Turkey and Yugoslavia accounted for about 14 per cent of total Western European coal consumption and for about 23 per cent of the total energy consumption; that is, coal played a relatively smaller role in the Mediterranean region of ECE than in the other regions. Particularly in Greece, Spain, Turkey and Yugoslavia coal will be given a much greater role in the forthcoming decade than hitherto, mainly as a base for electricity generation. In these countries demand is expected to double;

TABLE II.1 The West European Solid Fuel Balances 1973 and 1985

(hard coal and brown coal)

BALANCE	AUSTRIA				BELGIUM				FINLAND				FRANCE			
	1973 mill tce	1985 mill tce	change mill tce	change %	1973 mill tce	1985 mill tce	change mill tce	change %	1973 mill tce	1985 mill tce	change mill tce	change %	1973 mill tce	1985 mill tce	change mill tce	change %
1. PRODUCTION	1.8	1.2	-0.6	-33	8.8	7.2	-1.6	-18	-	-	-	-	27.8	-50	-13.7	12.9 to 15.4
(a) underground	...	...	...	...	8.8	7.2	-1.6	-18	-	-	-	-	26.6	-47	-12.5	12.9 to 11.4
(b) open-cast	...	...	...	...	-	-	-	-	-	-	-	-	1.2	-100	-1.2	-
+ 2. IMPORTS	4.3	5	+0.7	+16	7.3	9.1	+1.8	+25	3.7	7.8	+4.1	+111	16	+74	+12	24.1 to 31.6
% 3. EXPORTS	-	-	-	-	0.5	0.2	-0.3	-60	-	-	-	-	1.8	-14	-0.3	1.5
+ 4. Withdrawals from stock (% = additions)	-	-	-	-	0.6	...	...	...	0.2	...	...	...	2.3	...	...	...
= 5. TOTAL INLAND SUPPLIES	6.1	6.2	+0.1	+2	16.2	16.1	-0.1	- 1	3.9	7.8	+3.9	+100	44.3	-8	-3.8	35.5 to 45.5
of which to																
(a) power (heating) stations	1.3	1.1	-0.2	-15	2.1	6.3	+4.2	+200	1.8	4.5	+2.7	+150	15	-37	-5.5	8.5 to 10.5
(b) iron and steel ind. and coke ovens	3.6	3.9	+0.3	+8	9.9	8.6	-1.3	-13	0.8	1.9	+1.1	+100	15.3	+1	+0.2	15.5
(c) households and others	1.2	1.2	-	-	4.2	1.2	-3	-71	1.4	1.4	-	-	14	-18 to +39	-3.5	11.5 to 19.5
6. SHARE OF COAL IN TOTAL ENERGY DEMAND/SUPPLIES	22	13	-	-	27	9	-	-	9	11 to 12	-	-	18	-	-	12
7. SHARE OF NATIONAL COAL DEMAND/SUPPLIES IN WESTERN EUROPEAN TOTAL	1.5	1.4	-	-	4	3.6	-	-	1	2.4 b/	-	-	11	-	-	9.2

Sources used: for 1973: UN Statistical series J "World Energy Supplies", New York 1976 (Vol. 19); for 1985 Part I or most recent government plans available

Conversion factors: see UN Statistical series J "World Energy Supplies", New York 1976 (Vol. 19); tonnages are metric

b/ includes peat

- = nil or half the unit employed

... = not available

TABLE II.1 (cont.)

BALANCE	FEDERAL REPUBLIC OF GERMANY				GREECE				NETHERLANDS				NORWAY			
	1973 mill tce	1985 mill tce	change mill tce	change %	1973 mill tce	1985 mill tce	change mill tce	change %	1973 mill tce	1985 mill tce	change mill tce	change %	1973 mill tce	1985 mill tce	change mill tce	change %
1. PRODUCTION	136	134	-2	-1.5	4.3	14	+9.7	+225	1.8	–	-1.8	-100	0.4	1.2	+0.8	+200
(a) underground	104	100	-4	-4	–	–	–	–	1.8	–	-1.8	-100	0.4	1.2	+0.8	+200
(b) open-cast	32	34	+2	+6.3	4.3	14	+9.7	+225	–	–	–	–	–	–	–	–
+ 2. IMPORTS	10	...	...	...	0.7	...	...	...	3.9	6.5 to 7.5	+3	+ 80	1.2	0.6	-0.6	-30
% 3. EXPORTS	28	...	...	...	–	–	–	–	1.4	–*	-1.4	-100	0.1	0.6	+0.5	+500
+ 4. Withdrawals from stock (% = additions)	2	...	...	...	–	...	...	...	...	...	...	...	–	...	...	...
= 5. TOTAL INLAND SUPPLIES	120	...	...	...	5	15*	+10*	+200*	4.3	6.5 to 7.5	+2.7	+63	1.4	1.2	-0.2	-15
of which to (a) power (heating) stations	67	...	...	...	2.9	12.7*	+9.8*	+340*	0.9	3-4	+2.5	+290	–	–	–	–
(b) iron and steel ind. and coke ovens	27	...	...	...	0.3	0.5*	+0.2*	+66*	2.5	3*	+0.5	+20	0.7	0.6	-0.1	-14
(c) households and others	26	...	...	...	1.8	1.8*	–*	–*	0.9	0.5*	-0.4	-54	0.7	0.6	-0.1	-14
6. SHARE OF COAL IN TOTAL ENERGY DEMAND/SUPPLIES	31	22	–	–	30	...	–	–	7	5.8	–	–	6	4	–	–
7. SHARE OF NATIONAL COAL DEMAND/SUPPLIES IN WESTERN EUROPEAN TOTAL	30	...	–	–	1.3	3.4	–	–	1.1	1.6	–	–	0.3	0.3	–	–

Sources used: for 1973: UN Statistical series J "World Energy Supplies", New York 1976 (Vol. 19); for 1985 Part I or most recent government plans available

Conversion factors: see UN Statistical series J "World Energy Supplies", New York 1976 (Vol. 19), tonnages are metric

* = secretariat estimate

– = nil or half the unit employed

... = not available

TABLE II. 1 (cont.)

BALANCE	SPAIN				SWEDEN				TURKEY			
	1973 mill tce	1985 mill tce	change mill tce	change %	1973 mill tce	1985 mill tce	change mill tce	change %	1973 mill tce	1985 mill tce	change mill tce	change %
1. PRODUCTION	11.5	25	+13.5	+117	-	-	-	-	6	21	+15	+250
(a) underground	10	15*	+5*	+50	-	-	-	-	5*	6*	+1*	+20*
(b) open-cast	1.5	10*	+8.5*	+566*	-	-	-	-	1*	15*	+14*	+1500*
+ 2. IMPORTS	3.5	6.6	+3.1	+89	2.4	3.6	+1.2	+50	-	4	+4	+
% 3. EXPORTS	-	-	-	-	-	-	-	-	-	-	-	-
+ 4. Withdrawals from stocks (% = additions)	0.6	...	...	...	-0.1	...	...	...	...	...	...	...
= 5. TOTAL INLAND SUPPLIES of which to	15.6	31.6	+16	+103	2.2	3.6	+1.4	+64	6	25	+19	+316
(a) power (heating) stations	7.3	17.7	+10.4	+142	-	-	-	-	2	12	+11	+500
(b) iron and steel ind. and coke ovens	5.9	12	+6.1	+103	2.1*	3.4*	+1.3	+62	2	8	+6	+300
(c) households and others	1.8	1.9	+0.1	+6	0.1*	0.2*	+0.1	+100	2	5	+3	+150
6. SHARE OF COAL IN TOTAL ENERGY DEMAND/SUPPLIES	17	15	-	-	4.3	5.6	-	-	31	...	-	-
7. SHARE OF NATIONAL COAL/ DEMAND SUPPLIES IN WESTERN EUROPEAN TOTAL	4	7.1	-	-	0.6	0.8	-	-	1.5	5.6	-	-

Sources used: for 1973: UN Statistical series J "World Energy Supplies", New York 1976 (Vol. 19); for 1985 Part I or most recent government plans available

Conversion factors: see UN Statistical series J "World Energy Supplies", New York 1976 (Vol. 19), tonnages are metric

* = secretariat estimate

\- = nil or half the unit employed

... = not available

TABLE II.1 (cont.)

BALANCE	UNITED KINGDOM				YUGOSLAVIA				TOTAL WESTERN EUROPE[a]			
	1973 mill tce	1985 mill tce	change mill tce	change %	1973 mill tce	1985 mill tce	change mill tce	change %	1973 mill tce	1985 mill tce	change mill tce	change %
1. PRODUCTION	132	134	+2	+2	16.5	21*	+4.5*	+27	346	350*	+4*	+1
(a) underground	123	118	-5	-4	6*	4*	-2*	-33	284	240*	-44*	-15
(b) open-cast	9	15	+6	+67	10.5*	17*	+6.5*	+62	62	110*	+48*	+18
+ 2. IMPORTS	1.7	2-3	+0.8	+47	2.4	5*	+2.6*	+108	79	115 to 120*	+35 to 40*	+48
% 3. EXPORTS	2.7	2-3	-0.2	-7	0.3	-*	-0.3*	-100	34	25*	-9 to 10*	-26
+ 4. Withdrawals from stocks (% = additions)	...	...	...	...	-	...	-	-	8	...	...	...
= 5. TOTAL INLAND SUPPLIES of which to	131	135	+4	+3	18.6	25.8*	+7.2*	+39	400	440 to 445*	+40*	+11
(a) power (heating) stations	75	...	...	...	9.8*	17*	+7.2*	+73	195*	250*	+55*	+28
(b) iron and steel ind. and coke ovens	25	...	...	...	2.0*	3*	+1*	+50	110*	115*	+5*	+4
(c) households and others	31	...	...	...	6.8*	5.8*	-1*	-14	95*	75 to 80*	-15 to 20*	-18
6. SHARE OF COAL IN TOTAL ENERGY DEMAND/SUPPLIES	40	35	-	-	53	28	-	-	26	20	-	-
7. SHARE OF NATIONAL COAL/ DEMAND SUPPLIES IN WESTERN EUROPEAN TOTAL	32.8	30.5	-	-	4.7	5.8	-	-	100	100	-	-

Sources used: for 1973: UN Statistical series J "World Energy Supplies", New York 1976 (Vol. 19); for 1985 Part I or most recent government plans available

Conversion factors: see UN Statistical series J "World Energy Supplies", New York 1976 (Vol. 19); tonnages are metric

a/ including Turkey, non-reporting countries and peat

* = secretariat estimate

- = nil or half the unit employed

... = not available

the increase in demand will be about 50 mill. tce. and in production (mainly from brown coal and lignite open-cast mines) 43 mill. tce. Imports, mainly coking coal, will double too, and rise to about 10-12 mill. t. As a result of these new policies the Mediterranean region of ECE might account for 25 per cent of solid fuel consumption in Western Europe in 1985.

It is worthwhile mentioning that also in Northern Europe, hitherto considered peripheric in terms of coal consumption, the change in the coal policy has been substantial. In Norway, Finland and Sweden an additional demand of about 5-6 mill. tce. is expected, almost exclusively to be covered by imports. In Sweden, the increase is mainly due to the rapid development of the iron and steel industry. Additional solid fuel supplies are expected to an extent of 3 mill. tce. from the exploitation of Finnish peat reserves. In 1985 Scandinavia will account for 4 per cent of the total Western European coal consumption compared with 2 per cent in 1973.

Coal consumption in the traditional coal countries of North-West Europe as a group will not grow. Decreases are expected in Austria, France, the Federal Republic of Germany and Switzerland, whereas increases are expected in the United Kingdom and the Netherlands. Demand will remain stable in Belgium.

The interest in international co-operation will be certainly enhanced by the advent of the Southern and Northern European nations as significant producers and/or importers of coal. The share of these two regions in the total Western European coal consumption will increase from 16 per cent to 28 per cent between 1973 and 1985. Under these circumstances, the elaboration of a specific programme of transfer of scientific and technological knowledge, suited to the new needs of the Mediterranean region of the ECE would seem warranted. Such a programme might include the prospection of coal reserves; the transfer of open-cast mining technology; exchange of experience on combustion of low-grade coals in power stations; preservation of the environment; modernization of the infrastructure in particular at ports; joint R&D in the further validation of indigenous coal; training of managers, engineers, scientists, etc.

PART III

PERSPECTIVES OF THE COAL INDUSTRY IN NORTH AMERICA

A. THE MAJOR INFLUENCING FACTORS OF THE POST-WAR PERIOD

The role of coal in meeting North American energy needs is presently changing. The factors which have governed coal's fate in the past are seeking a new equilibrium. Without entering into detail, an attempt will be made in this section to identify the major influencing factors and the various stages of development of the coal industry, in order to better assess coal's future.

Allowing for some simplification, one might state that after 1945, North American coal went through three distinct periods:

- 1951-1961 during which coal production fell steadily from 536 mill. t 1/ to 388 mill. t;
- 1961-1970 during which coal production increased regularly from 388 mill. t to 565 mill. t; and
- 1970-1973 which was characterized by ups and downs of coal production.

The developments in the United States and in Canada followed the same pattern with the exception of the last phase in which the (small) Canadian coal production experienced a steady increase.

The major factors which have tailored these phases have been "domestic" in nature, i.e. did not result from a rise or fall in foreign coal demand. During these periods, North America had been a net exporter of coal; these net exports fluctuated between 6.3 per cent of production in 1951 and 9.6 per cent in 1970 (see Table III.1). These figures are too low to suggest a significant influence of foreign trade on the coal situation in general. However, these exports have been a significant factor for the economy of the coal exporting basins such as the Appalachian and Alberta/British Columbia.

The basic "domestic" factors have been on the one hand, changes in the growth rate of total energy demand (reflecting mainly the speed of economic growth) and on the other hand, changes of the comparative competitiveness of coal, oil, natural gas and nuclear power.

The influence of the business cycle and of economic growth on coal production had been notable. In periods of low growth rates of total energy demand such as 1951-1961 (plus 2.3 per cent annually) coal was not able to defend its position either in absolute or relative terms. In periods of comparatively high energy demand implying increased activities of the iron and steel and electric power industries such as the period from 1961 to 1970 (plus 4.8

1/ Metric unless otherwise stated.

per cent annually), the coal industry was able to redress its competitive position: coal recovered tonnage-wise but did not recover market shares. Finally, in periods of intermediary growth rates of energy demand such as 1970-1973 (3.6 per cent) the coal industry had to perform in a stop-go climate of uncertainty.

Thus, high energy demand growth rates smoothed inter-fuel substitution while low rates increased the pinch. Competition came primarily from natural gas and oil as can be seen from the Table III.1. The combined share of these hydrocarbons increased from 60 per cent in 1951 to 75 per cent in 1961. Further penetration of oil and gas was at a reduced speed. From 1961 to 1970, the increase was only from 75 per cent up to 78 per cent. This slow-down was the result of high over-all energy growth rates and of improved competitiveness of coal through mechanization, strip mining, unit trains, etc. During this period also, nuclear power entered competition both through the construction of power plants and through the projection of its brilliant future. Government policies have played a considerable role in shaping these trends and developments mainly by conducting nuclear research and development, regulating oil imports and domestic energy prices, and raising the standards of human and environmental protection.

B. THE NEED FOR A RE-ASSESSMENT OF COAL'S ROLE

However, this conjecture of comparative competitiveness of the major forms of primary energy during the 1960s changed at the beginning of the 1970s. The availability of natural gas began to dwindle. Deliveries from indigenous oil wells fell significantly. Nuclear power experienced difficulties to live up to earlier expectations. Coal's fresh impetus was checked by reduced productivity resulting from labour disputes and stringent health and safety regulations, by increased costs due to environmental legislation, by reduced markets for high-sulphur coals and by limited access to cheap open-cast coal fields. In 1976 it was estimated that only 49 per cent of United States coal production was in conformity with air pollution regulations. 1/

These developments called for a reappraisal of the role of the various forms of energy including coal, when the so-called oil crisis of 1973/1974 broke out. While this crisis gave a decisive impetus to undertake a complete and comprehensive reappraisal it would be wrong to neglect the other events mentioned and to concentrate attention on the impact of higher oil prices and trade disruptions.

However, the disruptions of oil supplies and the four-fold increase in oil prices during 1973/1974 added a new and strong motivation for a reappraisal of coal's future: indigenous coal emerged as a means of reducing reliance on foreign oil.

Attention focused in the first instance on a reassessment of "economically recoverable measured reserves" taking into account that the new energy price level has improved the viability of hitherto sub-marginal coal seams. While it is not possible to give an accurate and final picture since the definition

---

1/ United States Bureau of Mines, Effects of air quality requirements on coal supply, Washington, May 1976, page 5.

of what is economically recoverable remains subject to change and challenge, in particular in international comparisons, it is nevertheless evident that the tonnage of "economically recoverable measured coal reserves" in North America had been considerably upgraded between 1968 and 1975. These reserves which had been estimated in 1968 to be around 68 billion tons, were estimated to be 188 billion tons 1/ in 1974 and about 220 to 225 billion tons in 1975 2/. These reserves would be sufficient to sustain coal demand at the 1974 level for 425 years. These coal reserves correspond to about 90 per cent of fossil energy reserves in the region if oil shale and tar sands are excluded, or to 33 per cent if these are included.

These figures influenced strongly the scenarios of possible future energy supply and demand patterns which were elaborated shortly after the outbreak of the energy crisis. For example, the United States "Project Independence" held that coal's potential was not limited on the supply side, but on the demand side, i.e. coal was capable of satisfying almost any foreseeable demand by 1985 at prices near 1972/1973 levels; in its "accelerated supply scenario" designed to explore the possibilities of reducing or even eliminating oil imports, the coal production potential for the year 1990 was estimated at 2,803 million short tons compared with 599 million short tons in 1973 3/. A task force established by the provincial government of British Columbia where presently half the Canadian coal is mined, held that coal production could be increased in that province from the present 10 mill. t to as high as 100 mill. t in 1995 in order to meet domestic and foreign, including Western European coking coal requirements 4/. Exploiting coal's potential to such an extent would mean an annual growth rate of coal production of 10 per cent in the case of the United States and of 11 per cent in the case of Canada.

C. ANTICIPATED ABSOLUTE AND RELATIVE EXPANSION OF COAL DEMAND

How big the potential of coal may be: since coal's expansion is dependent upon demand, conditions have to be such that this demand materializes. It took the years 1974-1975 before a consensus appeared with respect to the desirable extent of such demand, the role to be played by coal in the total energy picture, the extent of government involvement and the mechanism through which the new policy would be translated into actual production and deliveries.

As regards the role assigned to coal, it does seem now that the objective of reducing dependence on foreign oil is a major but not an overriding objective. There are other problems which, as has been mentioned, have emerged already before the oil crisis and which called for a solution. It is in response to these needs that coal is being called upon at present to replace scarce

---

1/ B. Fettweiss, Contributions to the assessment of world coal resources, Leoben (Austria), 1975, p.45.

2/ First National City Bank, Energy Memo, October 1975, p.1. To the 217 billion tons mentioned for the United States, 6 billion tons have been added to reflect economically recoverable resources in Canada as quoted in 1/.

3/ Federal Energy Administration, Project Independence, Washington, November 1974, pp.106, 108.

4/ Coal Age, February 1976, p.23.

natural gas and indigenous oil in power generation, to meet demand from electric utilities which under earlier expectations would have been covered by nuclear power, and to increase exports in line with foreign coal demand.

In furthering these objectives the Governments of the United States and Canada have chosen to rely essentially on the market forces. However, they are anxious to create a general climate of confidence in the competitiveness of coal. They took or contemplate measures to re-establish a price relationship between the various competing forms of energy which reflect the world price situation or long-term scarcities. They further the long-term opportunities of coal in undertaking R&D in particular on coal gasification and liquefaction; the funds allocated to the 1976/1977 R&D budgets correspond to approximately 0.50 to 0.60 US$/ton of production 1/. In addition the United States Government has been authorized to order utilities to burn coal instead of oil; it can also grant subsidies to small enterprises which open up new low-sulphur underground coal mines; the United States Government also ended a moratorium on federal coal leases. These measures have certainly enhanced coal's competitiveness, but its future will continue to depend essentially on the hazards and uncertainties of the domestic and world energy markets, on the established stringent environmental protection standards, and on the capability of the industry to attract the necessary capital and labour, to increase productivity in mining and transport and to enhance the environmental acceptability of coal. None of these problems has been characterized as insoluble; the capital markets have been found able to fund the coal expansion; 2/ railroads are investing heavily in new hopper cars and unit trains 3/ new coal delivery systems are promoted and expansion programmes of the industry show a majority of projects in open-cast and low sulphur coal 4/. On the other hand, the shortage of skilled labour and conflicting interests of provincial, state or environmental authorities seem such that the actual expansion of coal will lie substantially below the potential rates.

Forecasters have become increasingly aware of these difficulties and of the inertia of the energy systems in the medium-term, say until 1985, and have in the course of the last two years, repeatedly downgraded coal production forecasts (for example, for the United States the original estimate of 1,100 million short tons as mentioned in "Project Independence" has been downgraded to 1,040 million short tons in the revised version of "Project Independence" and to 998 million short tons in the Bureau of Mines' revised energy study of December 1975).

1/ For Canada, see Mining Journal, 16 April 1976, p.330. For the United States, see Coal Age, February 1976, p.9.

2/ Coal Age, February 1976, p.81; see also 5/, p.60.

3/ European Coal Information Agency (Brussels), Bulletin No. 5/1976; (Canadian) Minister of Energy, Mines and Resources: An energy strategy for Canada: policies for self-reliance; Ottawa, p.28.

4/ Coal Age, February 1976.

The present study uses a medium-term growth rate of hard coal and brown coal production for North America as a whole of 4.5 per cent, since this rate has been suggested in official publications for the United States 1/ as well as for Canada 2/. This growth rate would bring North American coal production from 552 mill. tce in 1973 to about 940 mill. tce in 1985 (see Table III.1). Taking into account estimated net exports of 80 mill. t, consumption of hard coal and brown coal would rise from 524 mill. tce in 1973 to about 860 mill. tce in 1985, i.e. at a growth rate of 4.2 per cent.

From 1974 onwards, coal consumption is expected to grow faster than energy consumption, for which a growth rate of 3.3 per cent has been computed from a forecast of the Bureau of Mines of 3.2 per cent 3/ and of the Government of Canada of 3.5 per cent 4/. Accordingly coal's share in total energy consumption will grow from 19 per cent in 1973 to 22-23 per cent in 1985.

D. TRENDS OF INTERNAL CONSUMPTION AND EXPORTS

Long-established consumption trends are likely to prevail also in the medium run:

- deliveries will be concentrated on two sectors, power stations and the iron and steel industry which will absorb 90 per cent of internal coal supplies;
- the relative importance of power stations as users of coal will further increase.

As can be seen from Table III.1, 76 per cent of the internal deliveries (or 665 mill. t) are earmarked for consumption in power stations in 1985 (compared with 70 per cent or 368 mill. t in 1974). Fourteen per cent or 120 mill. t will serve the purposes of the iron and steel industry (compared with 18 per cent or 94 mill. t in 1974). Out of the remainder of the 1985 deliveries 25 mill. t go to United States gasification plants: this is a new market for coal, the future of which is considered most promising in the United States. For the year 2000 the Bureau of Mines foresees deliveries to gasification and liquefaction plants of as much as 355 mill. t which corresponds to 25 per cent of total internal coal demand 5/.

Net exports of coal have been estimated to reach 80 mill. t in 1985, which is

---

1/ Bureau of Mines, United States energy through the year 2000 (revised), Washington, December 1975, p.61.

2/ See Part I, Section D, Table I.5.

3/ See footnote 1 above, p.5 .

4/ (Canadian) Minister of Energy, Mines and Resources, An energy strategy for Canada: policies for self-reliance, Ottawa 1976, page 25.

5/ See footnote 1 above, Table 10.

an increase of 35 million tons against 1973 1/. The share of exports in total production (8.5 per cent) differs only slightly from pre-crisis figures (see Table III.1). The actual increase may differ substantially from those forecast in view of the uncertainties of the export markets and of government interventions, 2/ but also in view of the considerable potential for exports claimed by the coal exporting regions.

In view of the inertia of energy systems, the developments thus forecast seem to be realistic; while the estimated production growth would tap only forty per cent of coal's maximum growth potential, it requires nevertheless a considerable effort on the part of the coal mining industry, of the transport industry and of coal users. Two major uncertainties remain: the higher demand for coal has to be met concurrently with higher demand for environmental protection, and the possibility cannot be discarded that indigenous coal might be called upon to compensate for unforeseen (short term or lasting) shortfalls of other forms of energy.

1/ Bureau of Mines, United States energy through the year 2000 (revised), already quoted, table 33 estimates United States net exports of coal at 68 million metric tons; the Mining Journal, 9 May 1975, expects a doubling of Canadian exports of coal by 1985, while imports of coal are said to be likely to decline.

2/ (Canadian) Minister of Energy, Mines and Resources: An energy strategy for Canada: policies for self-reliance; already quoted, p.29: "The federal government therefore has taken the position that, before new export commitments for coal are undertaken, efforts should be made by the producers to determine whether a market exists in Canada for the additional coal production."

TABLE III.1 The North American Solid Fuel Balance
(hard coal and brown coal)

	1951	1961	1970	1973	1974	1985*/
1. Production, mill.tce	536	388	565	552	562	940
+2. Imports, mill.tce	25	12	18	16	18	...
-3. Exports, mill.tce	58	35	71	61	67	...
[4. Net exports, mill.tce]	33	23	54	45	50	80
[5. Net exports, in per cent of production]	6.3	5.9	9.6	8.2	8.9	8.5
-6. Additions to stocks, mill.tce	6	-7	16	-17	-8	...
-7. Bunkers, mill.tce	1	1	-	-	-	...
=8. Consumption, mill.tce	496	371	494	524	521	860
per cent	100	100	100	100	100	100
of which, by:						
(a) power stations, mill.tce	117	187	325	369	368	655
per cent	24	50	66	70	70	76
(b) coke ovens/iron and steel industry, mill.tce	119	80	101	98	94	120
per cent	24	22	20	19	18	14
(c) Others, mill.tce	260	104	68	57	59	85
per cent	52	28	14	11	12	10
9. Total energy consumption, mill.tce	1272	1599	2444	2719	2655	3780
of which share of:						
(a) solid fuels	39	23	20	19	20	22-23
(b) liquid fuels	39	44	43	45	45	45
(c) natural gas	21	31	35	33	33	19
(d) hydro-, nuclear power	2	2	2	3	3	15
10. Share of United States in total North American production of coal	97.1	97.9	97.8	96.7	96.7	96.7
11. Share of United States in total coal exports of the region	99.1	97.1	94.1	81.5	83.6	75-82

Sources: all actual data, except 8(a)-(c) have been taken from United Nations, World Energy Supplies 1950-1974, New York 1976; for 8(a)-(c) see OECD Basic Statistics of Energy 1951-1965, Paris 1967, and OECD Statistics of Energy 1960-1974, Paris 1975.

*/ = estimates; tons are metric; ... = not available.

PART IV

PERSPECTIVES OF THE COAL INDUSTRY IN EASTERN EUROPE INCLUDING THE USSR

A. SOME CHARACTERISTICS OF THE DEVELOPMENTS SINCE 1945

Aggregation of country data in groups such as "centrally-planned economies in the ECE region" 1/ are meaningless if they cover opposing trends. In the case of the east European energy economies, this danger of levelling out the problems does not exist. Certainly the starting point of the various countries after World War II was affected by differences in war damage, degree of industrialization, population sizes and densities, endowment in natural resources, etc. But the developments which took place after 1945 on the basis of these differences showed a remarkable unison:

- high but falling rates of growth of total energy demand;
- a slow substitution of coal by oil and natural gas, with coal remaining the primary form of energy No. 1;
- on balance, an insignificant if any improvement of specific energy consumption;
- growing importance of power stations as the main users of coal and of open-cast mining as the cheapest mining technology, etc.

This unison of trends and structural changes reflected a common attitude and a common or co-ordinated policy with regard to post-war reconstruction, economic growth through industrialization and mechanization of agriculture, urbanization, allocated housing space, electrification and dieselization of railways, private ownership of cars, preference for indigenous or regional energy resources, etc.

The unison of approach to energy and coal problems and the absence of abrupt breaks of trends or structures will also characterize the region for the period 1974-1990. Certainly changes will take place and they will be highlighted in the subsequent sections of this report. But they will take the form of changes in emphasis, of nuances, rather than of fundamental changes: the major physical growth indicators (as far as published) show only marginal annual modifications whereby the long-term developments (1980-1990) even counterbalance some of the changes introduced with respect to the medium term (1976-1980). Thus, as long as the purpose of the present analysis is to describe the direction of change and to identify the resulting consequences and problems for international co-operation, it seems justified to consider the centrally-planned economies of the ECE region as a group.

B. ANTICIPATED ENERGY DEMAND GROWTH TILL 1980

A feature of the post-war energy markets in Eastern Europe including the USSR

1/ Albania, Bulgaria, Czechoslovakia, German Democratic Republic, Hungary, Poland, Romania, USSR.

efficiency. The main reason is that the further growth of the economies in Eastern Europe cannot rely to the same extent as in the past, on the mobilization of new manpower resources. It has been estimated that for the region as a whole, the employment growth will fall from 1.7 per cent annually from 1950 to 1970, to 0.8 per cent from 1970 to 1990, due mainly to the flattening out of population growth, an unfavourable age structure, less participation of women, shorter working time, etc. As a consequence, the economic growth will depend on increased labour productivity, the substitution of labour by capital and new applications of science and technology, i.e. on a more capital-intensive type of production which generally requires a more energy-intensive type of production. In addition, investments in energy-intensive and converting industries and the infrastructure will, as in the past, affect the specific energy consumption adversely. On the other hand, there is a considerable potential for increased economy and efficiency of the order of one-third or more by the early 1990s provided supporting measures are taken by Governments. However, the scope for improvements in the short or medium term is estimated substantially below the target.

Judging from the official Five-Year plans and official energy demand forecasts for 1976 to 1980, the net results of the two opposing trends will be, at least in the short run, a deterioration of the energy input/NMP output ratio. As can be seen from Table IV.1, the NMP growth for that period will slow down by 15 per cent in comparison with the previous five-year period, while the growth rate for total energy demand falls by only 4 per cent.

C. THE ROLE OF COAL

Planners are thus faced with absolutely and relatively increasing demands for capital and labour from energy supply industries. Denying those demands would jeopardize the achievement of the overall economic growth target, satisfying them would reduce the margin of manoeuvre which was already narrow enough.

The solution envisaged will be a change in the growth pattern of the energy supply industries: the rate of growth of hydro-carbon production will be reduced from 7.8 per cent during 1970-1974 to 5.4 per cent during 1974-1980, and the growth rate of coal production will be stepped up from 1.7 per cent during 1970-1974 to 2.7 per cent during 1974-1980 (see Table IV.2). In other words coal will be given greater emphasis, the main motivation being that it was comparatively less affected by factor limitations:

- as regards capital needs - within the ECE region the coal industry is the least capital-intensive energy supply industry (which are all highly capital-intensive);
- as regards labour needs - the coal industry undertook to accelerate production without a change in the labour force, i.e. to increase labour productivity;
- as regards the resource base - coal accounts for 82 per cent of the economically recoverable resources of the region compared with 11 per cent for natural gas, 6 per cent for oil and 1 per cent for hydro power; natural gas and oil are experiencing growing geological and climatic difficulties impeding continued high growth rates; the attraction of foreign oil is reduced by high world market prices and by the desire of Governments to preserve a high degree of self-reliance.

has been the fact that increased energy demand was almost exclusively met by increased energy supply and that the other way of dealing with energy demand - increased efficiency of energy use - on balance has not played a significant role: energy consumption per unit of net material product (NMP) was around 0.7 during the post-war reconstruction period, deteriorated in the 1960s to reach about 0.8 and improved during the early 1970s to reach the initial level of 0.7. Thus, in order to produce one unit of national income, 0.7 units of energy were required in the early 1970s as well as in the 1950s.

The fact that the specific energy consumption did not change on balance does not mean that no attention was paid to efficient energy use. It means that two opposing trends on balance levelled each other out: on the one hand the trend of improved performance of the individual equipment and plant, on the other the trend towards using a greater proportion of converted energy (involving conversion losses), shifts in industrial structures towards energy-intensive industries (metallurgy, chemical industry, glass, ceramics, aluminium), greater transport needs, and a general emphasis on "extensive", rather than intensive growth.

Under these circumstances, the growing energy needs of the region had necessarily to be covered by a quantitative expansion of energy supply. This expansion had been extraordinary in the 1950s (with a growth rate of about 10 per cent annually) and, while falling, remained still above the world average in the 1960s and early 1970s.

TABLE IV.1 Actual Growth Rate of Net Material Product[1/], of Energy Demand and Specific Energy Consumption in Eastern Europe Including the USSR

	Annual growth rate of NMP	Annual growth of energy demand	Specific energy consumption
1951-1960	9.7%	7.1%	0.7
1961-1970	6.8%	5.5%	0.8
1971-1975	6.4%	4.7%	0.7
1976-1980	5.1-5.7%	4.5%	0.8-0.9

Sources: See UN Overall Economic Perspectives for the ECE Region up to 1990; in part: EC.AD(XIII)/R.3, Table 3 and EC.AD(XIII)/AC.1/R.1/Add.3; for energy demand: UN World Energy Supplies 1950-1974, New York, 1976.

In looking at the near future (Five-Year Plan 1976-1980), it seems that the forces tending to increase the specific energy requirements will, on balance, gain a greater weight than the opportunity for further economies and increased

1/ The difference between the GDP and NMP systems of national accounts results from the exclusion of non-material services from the latter concept. Non-material services include education, defence, housing, personal services etc.

The gain due to the acceleration of coal production (about 60 mill. tce) will not cover the new energy demands. As in other ECE regions, coal is not the unique answer to emerging energy problems, and certainly not at short notice. This is why in spite of the acceleration of coal production, the share of coal in the total energy picture of the centrally-planned economies of the ECE region will continue to fall and reach 40 per cent in 1980 compared with 44 per cent in 1974.

D. SALIENT FEATURES OF COAL DEMAND/SUPPLY DURING 1976-1980

The speeding up of coal's growth rate from 1.7 per cent during the early 1970s to 2.7 per cent during the late 1970s will not cause important changes in the production and supply structures:

- net exports from the region to outside destinations are estimated to grow further, although at a slower pace; this deceleration reflects constraints mentioned above (availability of capital and labour), as well as the expectation that intra-regional trade in coal will gain a greater momentum;
- hardly any change is foreseen in the relative importance of brown coal/lignite mining versus production of hard coal (on a ton of coal equivalent basis): the share of brown coal/lignite in total coal production will remain closely to 30 per cent; however, on a ton = ton basis brown coal/ lignite mining will expand at a rate which is two to three times higher than the rate of growth of hard coal;
- open-cast mining will continue to grow in relative terms (in 1980 about 41 per cent of the hard coal and brown coal produced (on a tce basis) will be mined in open casts compared with 39 per cent in 1974);
- internal deliveries: a considerably growing share of coal production will be consumed by power stations and district heating plants; industry, railways, households and small consumers will reduce the direct consumption of coal, but remain the second biggest outlet for coal; deliveries to the iron and steel industry, including coke ovens, will grow in absolute, but possibly not, in relative terms;
- there will be an increase in coal production in all countries (except Hungary), in particular in Poland and Romania.

E. PRODUCTION GROWTH BEYOND 1980

Little quantitative evidence is given by governments for 1980 and beyond, except for coal production which is expected to increase at a rate of 2 per cent from 1980 to 1990. This growth rate lies by one-quarter below the growth rate for 1976 to 1980. Increases are reported from all countries, Hungary, Romania and the USSR ranging about average (see Table IV.2).

In the government replies reproduced in Part I of the present study, no explanation was given for the deceleration of coal production growth during the 1980s. Whatever the underlying motivations or combination of motivations, the envisaged deceleration of coal production growth during the 1980s (unless revised) indicates to the international community outside that region:

TABLE IV.2 Solid Fuel Balance of the Centrally Planned Economies of the ECE Region, 1951-1990 (hard coal and brown coal)

BALANCE	ALBANIA						BULGARIA						CZECHOSLOVAKIA						
	1951	1959	1970	1973	1974	1980	1951	1959	1970	1973	1974	1980	1951	1959	1970	1973	1974	1980	1990
1. Production, mill.tce	-	0.1	0.3	0.4	0.4	0.8	3.3	7.4	14.8	13.6	12.3	19.6	36	57.3	77.3	76.5	77.4	88-90	91-93
+ 2. Imports, mill.tce	-	-	-	-	-	...	-	0.3	5.3	6.1	6.3	...	4.7	2.6	5.2	5.7	5.6	...	...
- 3. Exports, mill.tce	-	-	-	-	-	...	-	0.1	-	-	-	...	2.6	3.5	5.9	6.5	6.8	=	=
[4. Net exports, mill.tce]	-	-	-	-	-	...	-	-0.2	-5.3	-6.1	-6.3	...	-2.1	0.9	0.7	0.8	1.2	...	...
[5. Net exports, in per cent of production]	-	-	-	-	-	...	-	-2.7	-36	-45	-51	...	-6	1.6	0.9	1	1.6	...	...
- 6. Bunkers, additions to stocks, mill.tce	-	-	-	-	-	...	...	...	...	...	...	...	...	...	-0.1	-0.4	-0.1	...	...
= 7. Consumption, mill.tce of which by	-	0.1	0.3	0.4	0.4	0.8	3.3	7.7	20.1	19.7	18.6	...	38.1	56.5	76.7	76.1	76.3	82-84*	86-87*
(a) power stations, mill.tce	-	...	...	...	...	...	..	..	10.4	10.4	10.2	+	..	15.2	23.5	29.3	28.6	+	+
in per cent of (7)	-	...	...	...	...	...	..	..	52	53	55	+	..	27	31	38	37	+	+
(b) iron and steel industry (coke oven coke, 1/ brown coal, brown coal coke, converted into tce), mill tce	-	...	...	...	...	...	..	..	2.6	2.9	3.1	...	..	..	11.2	11.5	12	+	=
in per cent of (7)	-	...	...	...	...	...	..	..	13	15	17	...	..	..	15	15	16	%	%
8. Total energy consumption, mill.tce of which share of	0.1	0.5	1.4	1.7	1.8	2.6*	3.7	8.9	33.5	36.6	36.4	58	39.5	60.7	93.5	98.8	100.2	114-117	149-161
(a) solid fuels	37	33	24	25	26	31	90	87	60	54	51	%	96	93	82	77	76	68*	54*
(b) liquid fuels	54	53	62	57	58	49*	9	12	37	43	46	+	3	3	13	18	17	+	+
(c) natural gas	8	11	10	11	11	12	-	-	2	1	1	+	-	3	4	5	6	+	+
(d) hydro-, nuclear power	1	3	4	6	5	8*	1	1	1	2	2	+	-	-	1	1	1	+	++
9. Share of country in total coal production of the region	-	-	-	-	-	0.1	0.8	1.1	1.9	1.6	1.4	1.9	8.6	8.9	9.7	9.0	9.0	8.6	7.3
10. Share of country in																			
(a) total exports of coal	-	-	-	-	-	...	-	-	-	-	-	-	7.7	8.5	8.5	8.4	8.1	...	...
(b) total imports of coal	-	-	-	-	-	...	-	1.5	13.9	14.1	15.0	...	24.0	10.4	13.6	13.2	13.3	...	...
11. Share of brown coal/lignite in total coal production	100	100	100	100	100	100	94	94	97	97	98	98	50	56	64	64	64	66	72
12. Rate of growth of total coal production	-	+19.7	+7.0	+10.2	+4.7	+15	-	+10.7	+6.5	-2.9	-9.4	+8.1	-	+6	+2.8	-0.3	+1.1	+1.8	+0.3

TABLE IV.2 (cont.)

BALANCE	GERMAN DEMOCRATIC REP. 1951	1959	1970	1973	1974	1980	1990	HUNGARY 1951	1959	1970	1973	1974	1980	1990
1. Production, mill.tce	48.6	67.3	79.2	74.6	75	75-76	81	8.4	14	15.6	14.8	14.3	12-13	15-16
+ 2. Imports, mill.tce	5.5	12.2	12.2	12.7	11.5	...	...	0.9	2.3	3.4	2.8	2.9	...	...
- 3. Exports, mill.tce	1.4	4.3	2.6	1.9	2.2	...	...	0.4	-	0.1	0.1	0.1	...	...
[4. Net exports, mill.tce]	-4.1	-7.9	-9.6	-10.8	-9.3	...	...	-0.5	-2.3	-3.3	-2.7	-2.8	...	...
[5. Net exports, in per cent of production]	-8.4	-11.7	-12.1	-14.5	-12.4	...	...	+6	-16.4	-21	-18	-20	...	...
- 6. Bunkers, additions to stocks, mill.tce	...	...	...	...	...	...	...	...	-	-	-	...	...	...
= 7. Consumption, mill.tce of which by	52.7	75.1	88.8	85.4	84.2	85-88	...	9.0	16.3	19	17.5	17	14-16	17-20
(a) power stations, mill.tce	..	..	..	52.2	50.9	+	+	..	6.1*	7.4	9.4	9.1	...	...
in per cent of (7)	..	..	..	61	60	+	+	..	37	39	54	54	...	...
(b) iron and steel industry (coke oven coke, 1/ brown coal, brown coal coke, converted into tce), mill tce	..	..	2.6	2.8	2.7	...	...	..	2.2*	2.3	2.5	2,7	...	...
in per cent of (7)	..	..	3	3	3	...	...	..	13	12	14	16	...	...
8. Total energy consumption, mill.tce of which share of	53.1	76.4	103.2	114	119.2	142-146	+	10.6	19.3	32.6	36.5	37.2	48-50	75-78
(a) solid fuels	99	98	86	75	71	60	%	85	84	58	48	46	30-31	23-25
(b) liquid fuels	1	1	12	16	17	+	+	10	12	26	32	34	+	%
(c) natural gas	...	-	2	9	12	+	+	5	3	15	18	19		
(d) hydro-, nuclear power	-	-	-	-	-	+	+	-	-	1	2	2	+	++
9. Share of country in total coal production of the region	11.6	10.4	9.9	8.8	8.7	7.5	6.7	2	2.2	2.0	1.7	1.7	1.2	1.3
10. Share of country in														
(a) total exports of coal	4.2	10.5	3.7	2.5	2.6	...	...	1.2	-	-	-	-	-	-
(b) total imports of coal	28.1	48.9	32.0	29.3	27.3	...	...	4.6	9.2	8.9	6.5	6.9	...	...
11. Share of brown coal/lignite in total coal production	93	96	99	99	99	99	99-100	81	81	76	79	76	76*	80*
12. Rate of growth of total coal production	-	+4.1	+1.5	-2	+0.4	+0.1	+0.7	-	+6.6	+1	-1.7	-3.9	-2.2	+2.2

TABLE IV.2 (cont.)

BALANCE	POLAND							ROMANIA						
	1951	1959	1970	1973	1974	1980	1990	1951	1959	1970	1973	1974	1980	1990
1. Production, mill.tce	83.5	101.9	149.9	168.4	174	205-210	250-265	2.6	4.3	11.1	13	13.6	24.8	32-34
+ 2. Imports, mill.tce	0.1	1.8	1.3	1.2	1.2	...	...	0.1	0.8	2.9	4	4.2	...	...
- 3. Exports, mill.tce	27.8	19.8	32.0	39.6	44	+	...	-	-	-	-	-	-	-
[4. Net exports, mill.tce]	27.7	18	30.7	38.4	42.8	...	...	-0.1	-0.8	-2.9	-4	-4.2	...	...
[5. Net exports, in per cent of production]	33.2	17.7	20.4	22.8	24.5	...	...	-3.8	-18.6	-26.1	-30.8	-30.8	...	...
- 6. Bunkers, additions to stocks, mill.tce	0.5	0.3	-0.3	1.5	-0.9	...	...	...	...	...	...	...	...	...
= 7. Consumption, mill.tce of which by	55.3	83.5	119.6	128.5	132.1	150-165*	...	2.7	5.1	14	17	17.8	31*	...
(a) power stations, mill.tce	..	19.2	33.1	42.9	46.5	+	+	...	1.4*	6.2	8.3	10.8*	++	++
in per cent of (7)	..	23	28	33	35	...	...	...	27	44	49	61	+	+
(b) iron and steel industry (coke oven coke 1/, brown coal, brown coal coke, converted into tce)														
mill.tce	..	8.2	8.4	9.2	9.0	...	...	...	1.4	4.9	4.4	4.3	+	+
in per cent of (7)	..	10	7	7	7	...	...	...	27	35	26	24	%	%
8. Total energy consumption, mill.tce of which share of	56.5	87.3	138.9	153.6	157.9	190-200	290-300	10.9	23.3	61	74.2	74.5	96	+
(a) solid fuels	98	96	86	84	84	79-80*	%	25	22	23	23	24	32*	...
(b) liquid fuels	1	3	8	10	10	) +	) +	32	32	23	24	22	+	...
(c) natural gas	1	1	6	7	7	)	)	43	46	54	52	54	%	...
(d) hydro-, nuclear power	-	-	-	-	-	+	+	-	-	-	-	-	+	...
9. Share of country in total coal production of the region	19.8	15.7	18.8	19.8	20.0	21.3	21.2	0.6	0.7	1.4	1.5	1.6	2.5	2.7
10. Share of country in														
(a) total exports of coal	82.9	48.2	46.4	51.3	52.6	...	...	-	-	-	-	-	-	-
(b) total imports of coal	-	7.2	3.4	2.8	2.9	...	...	0.5	3.2	7.6	9.2	10	...	...
11. Share of brown coal/lignite in total coal production	2	3	7	7	7	7	12	18	27	42	45	48	62	62-63
12. Rate of growth of total coal production	-	+2.5	+3.6	+3.9	+3.3	+3.6	+1.7	-	+6.6	+9	+5.5	+4.9	+10.5	+2.9

TABLE IV.2 (cont.)

BALANCE	USSR[2/]							TOTAL						
	1951	1959	1970	1973	1974	1980	1990	1951	1959	1970	1973	1974	1980	1990
1. Production, mill.tce	238.3	393	450.4	489	494.1	567-582	700-710	420.7	645.4	798.6	850.4	861	990-1 020	1 200*-1 230*
+ 2. Imports, mill.tce	8.2	5	7.8	10.7	10.4	...	...	19.6	24.9	38.1	43.3	42.1	...	...
- 3. Exports, mill.tce	1.3	13.3	28.4	29	30.4	...	...	33.5	41.0	68.9	77.1	83.6	...	...
[4. Net exports, mill.tce]	-6.9	8.3	20.6	18.3	20.0	...	...	13.9	16.1	30.8	33.8	41.5	45-50*	...
[5. Net exports, in per cent of production]	-2.9	2.1	4.6	3.7	4	...	...	3.3	2.5	3.9	4	4.8	4.7*	...
- 6. Bunkers, additions to stocks, mill.tce	...	...	...	...	...	...	...	...	0.2	-0.5	1.1	-1	...	...
= 7. Consumption, mill.tce of which by	245.2	384.7	429.8	470.7	474.1	545*-560*	...	406.4	629	768.3	815.4	820.5	940-975*	...
(a) power stations, mill.tce	...	...	...	...	...	...	...	...	...	...	450*	...	...	...
in per cent of (7)	...	...	...	...	...	...	...	...	...	...	55	...	...	...
(b) iron and steel industry (coke oven coke 1/, brown coal, brown coal coke, converted into tce) mill.tce	...	51.4[3/*]	76.7[3/*]	83[3/*]	85.2[3/*]	...	...	...	78*	109*	116*	119*	...	...
in per cent of (7)	...	13	18	18	18	...	...	...	12	14	14	15	...	...
8. Total energy consumption, mill.tce of which share of	310.2	581.2	1 074.1	1 262.9	1 323.7	1 740*-1 760*	...	484.5	857.6	1 538	1 778.3	1 851	2 400-2 435*	...
(a) solid fuels	79	66	40	37	36	31-32*	...	84	73	50	46	44	40	...
(b) liquid fuels	18	25	34	36	37	+	...	13	19	28	31	32	+	...
(c) natural gas	3	8	25	25	26	=	...	3	7	21	22	23	=	...
(d) hydro-, nuclear power	-	1	1	1	1	+	...	-	1	1	1	1	+	...
9. Share of country in total coal production of the region	56.6	60.9	56.3	57.3	57.3	56.9	58.0	100	100	100	100	100	100	100
10. Share of country in														
(a) total exports of coal	3.9	32.4	41.2	37.6	36.3	...	...	100	100	100	100	100	100	100
(b) total imports of coal	41.8	20.1	20.5	24.7	24.7	...	...	100	100	100	100	100	100	100
11. Share of brown coal/lignite in total coal production	24	23	20	20	19	...	...	31	33	33	31	30	30	...
12. Rate of growth of total coal production	-	+6.5	+1.2	+2.8	+1	+2.5	+2.1	-	+5.5	+2	+2.1	+1.3	2.7	2

Footnote for TABLE IV.2

Symbols: ... = not available; - = nil or negligible; * = secretariat estimate; =, +, ++, % = stagnation, increase, rapid increase, decrease.

Conversions into coal equivalents: (a) of brown coal/lignite: Czechoslovakia 0.6; Albania, Bulgaria, Hungary, USSR 0.5; Romania 0.33; German Democratic Republic, Poland 0.3; (b) of hydro- and nuclear electricity: 1 kWh = 860 kcal or 1000 kWh = 0.123 tce.

Sources: (a) For 1951 to 1974 and all lines except 7a and b, see United Nations, World Energy Supplies 1951-1974, New York 1976; for line 7a see ECE Quarterly and Annual Bulletins of Coal Statistics for Europe; for line 7b see last mentioned source and ECE Quarterly Bulletin of Steel Statistics for Europe.

(b) for 1980 (1990) see Part 1 of the present studies and published Directives/Guidelines on the Five-Year Plans 1976-1980; in addition see for Albania, Bulletin d'Information de l'Agence, Télégraphique Albanaise, 25.7.1976, for Romania EC.AD(XIII)R.2/Add.1 and for the TOTAL see Petroleum Economist, June 1976.

(c) In order to establish a homegeneous time series from 1951 through 1980 (1990), the national data for 1980 (1990) were recalculated to conform with the (United Nations) methodology applied to the years 1951-1974 as follows: the base year of the national forecast (i.e. 1970 or 1974) was compared with the corresponding year in the United Nations series; the difference if any served as a yardstick to aligning the national forecast for 1980 (90) to United Nations methodology; fuel wood and shale oil were excluded, hydro/nuclear electricity was included at 1 kWh = 860 kcal.

1/ For fear of double-counting deliveries of hard coal for purposes other than coke manufacture have been excluded.

2/ Includes peat.

3/ Excludes brown coal coke; data given show published blast furnace coke use and have been increased by 15 per cent to approach total coke use.

- that coal exports to destinations abroad are not likely to grow substantially unless supported by financial and other commitments on the part of the interested coal importers;
- that the opportunity for a transfer of labour-saving and other new technologies is growing particularly with regard to long-distance transmission, gasification/ liquefaction, mining at greater depth both in underground and open-cast workings, process control and automation with the help of computers, automation also of labour-intensive auxiliary activities, etc.;
- that the experience which might be gained with stringent conservation measures in that region is likely to be most beneficial to other ECE member States.

Interests with regard to these questions being complementary, an intensified intergovernmental co-operation within ECE in these respects will be possible.

PART V

PERSPECTIVES OF THE COAL INDUSTRY IN THE ECE REGION AS A WHOLE AND IN THE WORLD

This part deals with the future role of coal in the ECE region as a whole, placing its prospects when possible in the broader context of the world coal economy.

A. A NEW POLICY FOR COAL

The coal industry of the ECE region has gone through various distinct phases in the course of its long history. After World War II reconstruction characterized the early 1950s. The ensuing period was one of increasingly fierce competition from oil. It is now evident that this period came to an end in 1973/74 as a result of the extraordinary increase in the price of oil. This event demonstrated the vulnerability of the energy economy of the ECE region as regards substantial price increases and supply disruptions and their negative effects on the balance-of-payment equilibria and economic growth. Measures became urgent to reduce the exposure of ECE countries to further price increases; if no measures were taken the European Economic Community would, in 1985, be facing a dependency on oil imports of the order of 64 per cent, 1/ and the United States of 23 per cent 2/.

No doubt, the oil crisis had been the major cause of the formulation of a new energy policy after 1973/74. But while it was the most powerful, it was not the only one; doubts developed about the availability of nuclear power, its competitiveness, safety and resource base. Considerable delays had occurred in the commissioning of nuclear fission plants as a result of growing stringent safety regulations and public resistance. Construction and raw material costs had increased under the impact of inflation, new regulations, unsatisfactory progress in standardization of equipment and low profitability of exploitation. The margin of competitive advantage was reduced and in certain cases even disappeared; the coming on the market of breeder and nuclear reactors on an industrial scale was again postponed and intermediate technologies - high temperature reactors - became necessary to reduce the strain on uranium resources which otherwise were likely to disappear before the end of the century. These were matters of considerable concern to governments. While some governments strengthened their involvement in nuclear programmes to overcome these difficulties and to become less dependent on oil, others began to reconsider their nuclear programmes and reduced them drastically. For these countries the balancing of energy supply and demand in 1985 and 2000 were additional problems to those caused by the oil crisis.

The above-mentioned reasons caused a reappraisal of the role of coal after 1973/1974. One reason was lacking, which had dominated coal policies in many

---

1/ Commission of the European Communities, Towards a new energy strategy, Brussels 1974, p.15.

2/ Federal Energy Administration, Project Independence, Washington 1974, p.28.

western European countries during the 1960s: employment and regional development policies. The tightness of the labour markets during the years immediately preceding and following the oil crisis had rendered such policies less urgent. Also, another reason - earning foreign currency through coal exports - quite powerful in the past, seemed of less importance during the formulation of the new coal policy after 1973/74. Less easily-available investment funds and a desire to export high-value products based on coal (electricity, synthetics), rather than coal, seemed to be at the basis of this somewhat more hesitant attitude.

These changes in emphasis and the new reasons were so important that they necessitated and, indeed, prompted the formulation of an altogether new coal policy in the ECE region during 1974 and 1975. However, born out of a crisis and drafted in a situation of uncertainty these policies have a strong short- and medium-term connotation. They would be subject to further change, as the future of the world energy economy becomes clearer and as constraints affecting coal mining and use are lifted. The following analysis will, therefore, not only describe and assess the policies formulated in 1974 or 1975, but also indicate the margin of manoeuvre which governments have or could gain in future with respect to coal, and which may be the basis of a further adaptation of their coal policy.

B. THE DETERMINANT FACTORS

(a) Competitiveness

The competitive position of coal has improved since 1973/1974. A substantial competitive advantage over oil was gained in the power station market, particularly in the United Kingdom and the United States. Sub-marginal coal fields or seams have become viable under the impact of the four-fold increase in crude oil prices; losses (and hence the burden on public budgets) were reduced. However, since 1974 a considerable portion of the competitive edge gained over oil as a result of the oil price increases has been lost. Higher coal prices had been introduced in the centrally-planned economies of the ECE region to induce industrial consumers to greater economy. In the market economies, the substantial rise in coal prices was motivated partly by increased labour costs (accounting for up to 50 per cent of production costs) and inflation and partly by the desire to reduce loss and government assistance. Coal's competitive advantage was further reduced by the greater ability of the oil industry to optimize its operation over a range of countries and limit price increases in particular markets in line with expected competition from coal. These various interactions are best illustrated by the developments in the Federal Republic of Germany, a country in which world market changes are rather easily felt: in this country crude oil prices went up by 309 per cent during 1970-1976, while the price of heavy fuel oil which competes with coal in the power generation market, went up by 145 per cent only; the price of indigenous coal rose by 92 to 95 per cent in the same period 1/. In general, the short-term price formation on the ECE energy market resulted in a smaller improvement of coal's competitive position than expected immediately after the four-fold increase in crude oil prices in 1973/1974.

1/ Handelsblatt, 21.12.1976.

(b) The Resource Base

Turning to the geological parameters, it is necessary to begin with a warning. It is striking to note how incomplete and uncertain our knowledge of the structure, location and volume of energy reserves is. This uncertainty is mainly due to the fact that the exploration effort, its intensity and orientation, depends on demand, more exactly: on the demand expected to materialize in ten or twenty years time. In addition, the data reflect the specific technical and economic conditions prevailing at the time and place of the exploration; they may even be biased by the commercial interests of those who release them to the public. The international comparability of these data is hampered by differences in projected demand and demand patterns, differences in technical know-how and economic systems, in particular with regard to the allocation of costs. Accordingly, notions such as "reserves" and "resources" and definitions of what is "economically recoverable" whilst commonly used, contain indeed a different message. All these reservations do not invalidate the data provided by institutions such as the World Energy Conference (which are used here) or by eminent specialists 1/. They should render the analyst conscious of the fact that these data provide only orders of magnitude whose validity depends on assumptions made in the 1960s and even before, assumptions which have been marginally modified in the last few years in the light of new technological and demand trends. It is worthwhile underlining that the figures do not show the ultimate resources which may be virtually inexhaustible, but the proven measured reserves which are considered recoverable under the economic and technical conditions assumed to prevail in the next two or three decades.

Availabilities. All the regions and major coal-producing countries of the ECE area are endowed with huge reserves of economically-recoverable coal. Seventy-eight per cent of the world coal reserves are located in the ECE area (it being understood that exploration elsewhere in the world has not been as intensive as in the ECE area). As far as the ECE region as a whole is concerned, these reserves would last 230 years at the 1973 production level and 75 years at a production growing at the rates established after 1973/1974 (see Table V.1). Even at double these growth rates the reserves would prove sufficient to cover demand for 30 to 60 years in the major coal-producing countries of the ECE region.

1/ Such as Günter B. Fettweiss, Weltkohlenvorräte, Essen 1976.

TABLE V.1 The Economically Recoverable Coal Reserves of the ECE Area

(absolute, and relative to production)

Region/country	Economically recoverable reserves 10^9 t	Depleted at 1973 production in ... years	Depleted at 1973-1985 production growth rates in ... years
Total Western Europe, of which:	70	154	154
Fed.Rep. of Germany	40	185	185
United Kingdom	45	345	345
Total Eastern Europe, of which:	232	178	69
Poland	57	288	74
USSR	137	204	76
Total North America, of which:	187	332	62
United States	182	333	62
ECE	530	230	73
World	666	215	59

Source: World Energy Conference, Survey of energy resources 1974, New York 1974, pp.52 and 58; for Poland: Colliery Guardian, April 1975, pp.126-127; for United Kingdom: Government information.

Only Finland, Iceland, Denmark, Ireland, Malta, Cyprus and Switzerland have no coal reserves. Can it thus be concluded that the endowment of the ECE region as a whole with coal is not a limiting factor to coal's enhanced role?

The answer to this question does not depend solely on the ratio between the economically-recoverable reserves and production. It also depends on the characteristics of these reserves, of their location, on their chemical/physical properties, on their suitability for economical mining techniques, etc. Some of these characteristics are looked into below.

Location. One factor which will play a growing and limiting role already in the medium term, is the growing distance between coal producing and consuming areas. It can be estimated that between 1973 and 1985 76 per cent of the net increase in coal production will be in "distant" coal basins (see Table V.2).

TABLE V.2 Increased Coal Production During 1973 and 1985

Country	Total increase mill tce	of which from "distant" basins	
		mill tce	basins
USSR	150	115*	Ekisbastuz, Kusnetsk, Kansk-Achinsk, Karaganda
Poland	60	25	Lublin
United States	365	290	North Great Plains, Rockies
Canada	20	20	Alberta, Saskatchewan, British Columbia
Greece	10	10	Ptolemais, Amyntaion
Turkey	15	...	Elbistan
Above countries	605	460	(76 per cent of total production growth)
Total ECE	657-692	...	

* = Secretariat estimate

The growing distance between the new mines and industrial or densely populated consumption areas is not a phenomenon restricted to a few countries such as the United States, Canada or the USSR. It is a general feature also observed in countries where distances do not play an important role, for example, in France (growing share of production from the Lorraine basin), Belgium (growing share of production from the Campine basin), the German Democratic Republic (growing share of the Cottbus coal field), the United Kingdom (the new Selby development) and Poland (the new Lublin coal field). Accordingly, the map of coal production centres will undergo a significant change: whereas at the beginning of this century, coal producing and industrial regions coincided, separation will become a current feature during the 1980s and beyond. The "distant" coal fields which accounted for 17 per cent of coal production in 1973, will account for 27 per cent in 1985 - a formidable challenge on coal transportation and conversion, and an indicator for the geological constraints under which the expansion of coal in the "old" coal areas would have to take place.

Properties. Another limiting factor might be the discrepancy between the available and the required chemical/physical properties of coal, particularly between the availability of and demand for high calorie coal, coking coal, low sulphur coal and anthracite.

The statistical evidence for the availability in the ECE region of prime coking and low sulphur coals is incomplete. Not all countries report on the properties of their coal reserves; ranking standards differ between countries; average indications sometimes have only little value since properties may change even within a given deposit. Finally, the identification of one particular parameter - however useful this might be - does not characterize the coal sufficiently from the point of view of its suitability for a given use, since unsatisfactory properties may be overcome during preparation and blending.

Any quantitative assessment of the qualities and properties of economically-recoverable coal reserves can therefore serve only as a rough approximation.

Table V.3, although incomplete, shows that in the ECE region as a whole, its sub-regions and its major coal-producing nations, the economically-recoverable reserves of high calorie coal, coking coal and low sulphur coal (less than 1 per cent sulphur content by weight) are abundant and do not constitute a limiting factor for coal supply growth. For ECE as a whole, 70 per cent of the economically-recoverable reserves are above 5,700 kcal/kg. Approximately 20 per cent is said to be of coking quality. Hard coals with less than 1 per cent sulphur content represent about 10 per cent of the reserves in Western Europe and 15 per cent of the reserves in North America; low sulphur brown coal/lignite holds an even bigger share: more than 33 per cent of the reserves in Western Europe and more than 90 per cent in North America are low sulphur.

But problems may arise in smaller countries or having a less favourable geological spread. Out of the 22 ECE countries having coal reserves considered economically-recoverable, only 16 dispose of high calorie (hard) coal reserves and only 12 possess coal reserves of coking quality (the bulk of which - 93 per cent - is located in four countries: Federal Republic of Germany, Poland, United States, USSR). As regards low sulphur coals, they seem to be lacking only in Yugoslavia, Portugal and Norway.

Depth. Reserves are considered at present economically recoverable if, inter alia, the maximum depth is not more than 1,200 m for hard coal and not more than 500 m for brown coal/lignite. Obviously, the nearer a coal deposit lies to the surface, the greater its viability.

Table V.3 shows an encouragingly high proportion of coal reserves suitable for open-cast mining: in Western and Eastern Europe 80 per cent and more of the economically-recoverable reserves of brown coal/lignite are suitable to open-cast mining; in North America the percentage is lower, 30 per cent or so. As concerns hard coal reserves, 24 per cent in North America and 1 per cent in Western Europe are suitable for open-cast mining; the Western European ratio may also be representative for Eastern Europe, except in the USSR where huge reserves of hard coal are suitable for open-cast mining.

Other characteristics. The distance between the deposit and the final consumer, the calorific value, coking and caking property, sulphur content and depth of the coal are not the only parameters which have to be taken into account to assess the constraints which reserves may impose on supply growth. Thickness/inclination/uniformity of seams, ash/volatile matter/salt content, the occurrence of gas/heat/water, geomechanics play a role which cannot be quantified. All these factors interact with each other and do this increasingly in a restricted sense since the most suitable deposits are already being mined, sometimes for several decades.

Conclusion. Are coal reserves a limiting factor to coal production? A conclusion seems now possible: economically-recoverable reserves are abundant in the ECE region as a whole, in its sub-regions and in the major coal-producing countries. This holds both with respect to volume and property (calorific value, coking property, low sulphur content). Constraints appear in smaller countries due to the uneven spread of the various types of coal. Constraints appear also as a result of the long-lasting exploitation effort and the worsening of geological conditions: deeper, thinner, less uniform and more inclined seams have to be worked. Under these conditions, the development of

TABLE V.3 Economically-Recoverable Reserves of Coal, of Premium Coal and of Coal Suitable for Opencast Mining
(in million t)

Region/country	High calorie = hc Low calorie = bc	Total econ. rec. reserves	Of which of coking quality	Coals with Sulphur content		Reserves suitable for opencast mining
				below 1%	below 3%	
Western Europe						
Austria	bc	64	–	28	64	5
Belgium	hc	–	–	–	–	–
Fed. Rep. of Germany	hc	30,000	18,000	...	30,000	–
	bc	9,571	–	9,460	9,571	9,571
France	hc	443	349	443	443	–
	bc	15	–	–	15	4
Greece	bc	...	...	...	...	...
Italy (mainland)	bc	33	–	33	33	33
Netherlands	hc	1,843	...	1,843	...	–
Norway	hc	2	2	–	2	–
Portugal	bc	25	–	–	25	23
Spain	hc	907	228	354	907	–
	bc	736	–	15	41	347
Sweden	hc	30	–	30	30	–
Turkey	hc	134	80	134	134	–
	bc	...	–	–	...	...
United Kingdom c/	hc	3,871	1,635	296	3,871	298
Yugoslavia	hc	70	–	–	–	–
	bc	16,800	–	–	16,800	12,540
Total above	hc	37,300	20,300	3,100	35,400	300
	bc	27,250	–	9,500	26,500	22,500
Eastern Europe						
Albania	bc	...	–	...	...	...
Bulgaria	hc	29	–	–	29	–
	bc	4,358	–	–	346	4,012
Czechoslovakia	hc	11,600 a/)	10.5%	...	...	...
	bc	12,500 a/)		...	...	...
German Dem. Rep.	hc	...	...	...	...	...
	bc	25,300	–	...	...	25,300
Poland	hc	50,000 b/	15,000	...	...	–
	bc	5,000 8,000 b/	–	...	...	5,000 8,000 b/
Romania	hc	50	...	...	...	...
	bc	1,100	–	...	...	...
USSR	hc) bc)	136,600	22%	...	...	...
Total above	hc	150,000 */)	45-50,000 */	...	...	...
	bc	80,000 */)		...	...	...
North America						
Canada	hc	5,080	...	5,006	5,073	3,621
	bc	457	–	457	457	457
USA	hc	158,725	39,307	16,930	...	35,322
	bc	23,056	–	21,442	...	6,363
ECE, total	hc	360,000 */)	110,000 */	...	...	...
	bc	130,000 */)		...	...	...

Source: World Energy Conference, op. cit., pp. 50-60, 307-321.

a/ "Proven Accessible Reserves".

b/ Colliery Guardian, April 1975, pp. 126-127.

c/ Operating reserves only, accessible to existing collieries and opencast mines.

*/ Note: For some reserves, coals with less than 1 per cent sulphur are reported only together with coals up to 3 per cent of sulphur content; figures of low sulphur coals are, therefore, understated

D

virgin coalfields at greater distance, possibly by open-cast mining, becomes the main lever to make the virtually abundant coal reserves actually and economically viable. The other lever - "new" mining techniques such as hydraulic and chemical mining or underground gasification - is quantitatively insignificant or still at the pilot stage. With these reservations, economically-recoverable coal reserves in the ECE area and its major sub-regions are sufficient to sustain a growth of coal demand which could go substantially beyond the presently envisaged rate without undue prejudice to the demands of future generations.

(c) The Labour Force

Coal mining is a labour-intensive industry, particularly underground mining. In 1970 about 2.2 million men were on the books of the ECE coal mining industry (underground and open-cast). Manpower costs accounted for 50 per cent of total costs of underground production. To what extent will the availability of manpower and the development of labour productivity restrain coal production growth during 1973-1985?

Employment growth. It could be argued that the availability of manpower should generally not be a limiting factor since the labour force employed in coal mining does not constitute more than 0.4 per cent of the active population (15 to 59 years of age) in the ECE region (Table V.4), the highest share being not more than 1 per cent (in Poland). If the appropriate incentives and the necessary mobility of labour were ensured, the coal industry should be capable of attracting the required labour force, if necessary even to the detriment of other employers.

TABLE V.4 Labour Force, Total and in Coal Mining

Region	active population, 1970 (15-59 years) in 000	of which employed in coal mining (underground and open-cast), 1970		Projected growth rate of active 1970-1985
		000	%	
Western Europe	205 800	520 a/	0.3	0.8
Eastern Europe including USSR	206 800	1 500	0.7	1.1
North America	130 700	165 b/	0.1	1.3
ECE	543 300	2 200	0.4	1.0

Source: United Nations, Economic Survey of Europe in 1974, Part II, Table VII-12.

a/ Of which 489 underground.

b/ In 1973.

However, problems may arise caused by the fact that the active population, while still growing, will do so only at half the rate of the recent years. Thus, the coal industry might have to compete for new labour in a much tighter labour market. All other factors remaining equal, the growth of active population could sustain a rate of coal production of only 0.7 per cent in the coal-producing countries of Western Europe during 1970-85, of 0.8 per cent in Eastern Europe, excluding the USSR, of 1.2 per cent in the USSR and of 1.3 per cent in North America. A higher coal production growth would require greater internal and external mobility, and/or a shift of occupational patterns in favour of coal.

Under European conditions, these pre-requisites are not likely to be fulfilled: internal mobility is very limited; migration seems subject to increasingly restricted policy, even in boom periods; the recruitment of young labour (15 to 19 years) may pose in some countries problems due to demographic factors, increased length of education and the growing role of "distant" coal fields. It is therefore not surprising that European governments generally expect no increase in employment in the coal industry. On the contrary: Western European coal producers expect continued absolute decreases in the number of men on the books and East European coal producers consider a stagnation more realistic than an expansion.

Under North American conditions, the recruitment of new labour seems less critical. The projected growth of active population is higher than in Europe and the mobility of labour is traditionally high. The increase in the labour force in the United States coal mines, corresponding to an expansion of coal production to around 940 mill. tce would be 36,000 men; problems could arise locally and temporarily if production were to double in comparison with the presently envisaged tonnage, in which case 177,000 men would have to be added to the labour force by 1985 1/.

Labour productivity. In view of the poor prospects for employment growth, most European coal-producing countries have to raise labour productivity to raise coal production. During 1960 to 1975, output per man and year underground increased at an average rate of 3.7 per cent in Western Europe, 3.3. per cent in the USSR and 2.7 per cent in Eastern Europe, excluding the USSR. Since the early 1970s, productivity growth rates have become erratic and tend to slow down. The level of underground productivity reached in 1973 was fairly similar: 706 tons per man and year underground in Western Europe, 674 tons in Eastern Europe, excluding the USSR, and 600 tons in the USSR.

In the United States, productivity trends were different from Europe. As a result of labour disputes and new safety and health regulations, productivity fell during 1967-75 on the average by 4.9 per cent annually. The level recorded in 1973 was still high in comparison with European standards: around 2,600 tons per man and year underground 2/.

The pre-requisites for a return to earlier performance records in Europe and the United States are manifold ranging from stable labour relations towards a

---

1/ Federal Energy Administration, Project Independence, Report of the Task Force Coal, Washington 1974, pps. 18 and 34.

2/ Leslie C. Gates, USA's coal industry stagnates and productivity continues to decline, World Coal, November 1976, p.38.

favourable ratio between ("unproductive") development work, and actual extraction. The measures envisaged by industry and supported by governments include:

- a switch to less selective and less labour-intensive mining techniques: open-cast, in situ gasification, hydraulic and chemical mining;
- a refinement of existing technologies through complex mechanization, automation and remote control;
- the adaptation of mining management and operations, in particular the concentration of workings, and the combination of operations, for example in "coal-plexes";
- the greater efficiency of coal extraction through the exploitation of thin seams, inclined seams, seams beyond 2 m thickness, safety pillars, and through the replacement of room-and-pillar working by short and longwall mining;
- the orientation of R&D towards new mining methods such as in-situ gasification, solvent extraction, continuous haulage, greater attention paid to mining at greater depth or off-shore, application of new technologies to mining such as interface sensing devices;
- extended safety and training programmes.

What will be the yield of these productivity measures by 1985? What are the upper limits? No evaluation of this kind has been explicitly made in the national reports, probably in view of the uncertainty and complexity of the subject matter. However, remarks made and other information available with regard to production and employment growth, permit to draw a conclusion with respect to the future development of "output per man and year". This indicator is a very rough approximation for productivity in coal mining. It has the advantage of any macro-indicator of giving a first overall insight, but it has also the traditional disadvantage of aggregating elements of greater or lesser relevance or of doubtful international comparability. Table V.5 suggests that during 1973-85, the potential for annual productivity gains in underground mining would be around 3.5 per cent in Western Europe, 3.3 per cent in Eastern Europe, including the USSR and 3.1 per cent in the United States, if production and employment grow at the expected rates.

TABLE V.5 Projected Growth of Labour Productivity in Coal Mining, 1973-1985

in %

Region	Annual production growth	Annual employment growth	Annual productivity growth	Pro memoria: annual productivity growth 1960-1975
Western Europe (underground)	+ 0.6	- 2.9 a/	+ 3.5	3.7
Eastern Europe including USSR (underground)	+ 2.4	- 0.7 a/	+ 3.3	3.1
USA (total)	+ 4.5	+ 1.4 b/	+ 3.1	- 4.9 c/

a/ 1961-1973 historical rate.

b/ Federal Energy Administration, Project Independence, Report of the Task Force Coal, Washington 1974, p.46.

c/ In 1967-1975.

In conclusion: the demographic and social factors affecting the growth of population, in particular the active population will constitute a limiting factor to coal production growth in Europe, which could only be overcome by a rise of labour productivity which itself seems to be restricted to 3.3 to 3.5 per cent annually between 1973 and 1985. In North America, population growth and mobility are such that the availability of manpower does not constitute a limiting factor, but United States production targets for 1985 could not be achieved without a reversal of the decline of labour productivity.

(d) Capital Needs and Sources

The energy industry is a capital intensive industry. Any expansion of energy production, conversion, transportation and use entails a considerable demand for capital. It is to be feared that future demands for capital will be even bigger: easily exploitable energy resources are becoming scarce; environmental considerations result in additional costs; the more capital-intensive forms of energy, particularly electricity and nuclear power, are gaining a higher share in the supply pattern; inflation escalates construction costs; savings from greater efficiency and greater unit sizes grow at a lower pace than in the past; and last, but not the least, the policy pursued by ECE governments since 1973-74 to promote energy production from indigenous or sub-regional resources, will place an additional strain on available funds 1/. Accordingly, ECE

1/ P.B. Frankel estimates that in comparison with Middle East crude oil, investments in United States surface mined coal would cost three-and-a-half times as much, deep mined coal would cost six times as much, North Sea oil twelve times, oil sands synthetic twenty times, oil shale synthetic twenty-five times, coal gas synthetic twenty-five times, and coal hydrocarbons thirty five times; nuclear power would be far more expansive still (The Petroleum Economist, February 1975, p.59).

governments anticipated an increased investment per unit of output and, with the possible exception of the United States 1/, a growing share of investments in the energy economy relative to total gross fixed capital formation.

Also coal mining and preparation are capital-intensive activities, and investments per ton of output from open-cast or from underground mines grow under the impact of a deterioration of geological conditions, the substitution of capital for labour, the growing sophistication of equipment, the higher degree of mechanization and automation, and escalating construction costs. The capital necessary for supporting the anticipated growth of coal production between 1973 and 1985 has been estimated at around 20 . 10^9 US$ of 1975 value for North America and 13 . 10^9 unit of accounts for the European Economic Community 2/. No figures are available for Eastern Europe and the USSR, but have been estimated at nearly 17 . 10^9 US$ 3/. This would bring the total in ECE capital needs for coal expansion up to 50 . 10^9 US$ or 71$ per ton of new capacity.

These impressing absolute figures appear less in relative terms: investments in coal production and preparation during 1975-1985 would amount to only 0.1 per cent of total energy investments in Canada, 3 per cent in the United States and 5 per cent in the European Economic Community, compared with 19 per cent for oil and 28 per cent for nuclear power 2/ 3/. Some countries, particularly the United States, Canada, Poland, Romania and the USSR have the possibility of reducing the financial burden by increasing the share of coal production from open-cast mining, for which investments per ton may be only half those for underground mines 4/.

The fact that coal requires less capital per ton produced than any other fossil form of energy within the ECE region, renders financing problems less stringent. Accordingly, no major concerns have been expressed by those governments who addressed this issue, although the extraordinary capital needs of the energy economy as a whole, including transportation and conversion of coal, may have a side effect on the coal industry's capability of attracting the necessary capital. In countries with capital markets, the share of funds supplied by public authorities may therefore tend to rise. However, such an increase would not seem to reflect a shortage of capital, but the desire, on the part of governments, to control any rise of the price of coal or of the influx of foreign ownership. This being stated, it is also obvious that investment

---

1/ Federal Energy Administration, National Energy Outlook, Washington, February 1976, p.43.

2/ For the United States, Federal Energy Administration, National energy outlook, Washington 1976, p.306 (17.7 x 10^9 US$ at 1975 value); for Canada: Ministry of Energy, Mines & Resources, An energy strategy for Canada - policies for self reliance - summary; Ottawa, 1976, p.18 (2.3 x 10^9, Canadian dollars at 1975 value); for the European Economic Community, see 3/.

3/ United Nations, Status and prospects of coal production and utilization in developed and developing countries, New York 1977, p.25 (E/C.7/67).

4/ Ratio derived from a comparison of two model mines in the United States, a deep mine with an annual production of 1 mill. short t from a 1.83 m thick seam at a depth of approximately 215 m, and a strip mine with an annual production of 1 mill. short t, from a 1.83 m thick seam and a 10:1 overburden ratio (Federal Energy Administration, National energy outlook, op. cit. p.306).

funds are not plentiful and that priorities have been established in favour of investments serving domestic needs. A rise of coal production in response to foreign demand would increasingly require a financial commitment on the part of the importer.

(e) Environmental Constraints

The production, conversion and use of all forms of energy affects, at varying degrees, the human environment. Also, the activities of the coal industry, in spite of considerable progress in the control of subsidence, noise and dust can still pose important environmental problems, particularly through the release of heat, carbon monoxide and dioxide, nitrogen oxide and sulphur dioxide from coal-fired power stations, and release of toxic effluents from coking plants and mines, and the destruction of the flora and fauna of open-cast mining.

The means at the disposal of the coal industry to reduce the environmental consequences of its activities are generally considered compatible with the needs 1/. They comprise new technology (fluidized-bed combustion and de-sulphurization, hydrogenation, in-situ gasification, improved preparation techniques, recultivation) and operational changes (preferential production of low sulphur coals, use of waste heat for district heating, etc.). Constraints may appear locally as the result of specifically stringent air/water/reclamation standards, or temporarily as a result of the time needed to adapt to production and conversion techniques. More serious problems may appear in arid areas if coal slurry pipelines are envisaged for the long-distance transport of coal, and in the long term, if the combustion of coal (as the combustion of any other fossil fuel) prejudices the CO_2 balance of the globe.

The proportionate increase in coal production costs can be substantial, depending on local conditions, the characteristic of the deposit, the properties of the coal and the technologies applied 2/. For example, the recultivation of excavated areas in the United Kingdom has been estimated to account for 4 per cent of production costs, and for up to 8 per cent of production costs in the United States depending mainly on seam thickness. The cost of environmental control in coal gasification and liquefaction is estimated at about 10 to 20 per cent of total gasification/liquefaction costs. The reduction of the sulphur content of high-sulphur coals from deep mines, by 40 to 60 per cent is estimated to cost around 5 to 6 per cent of total production costs of high-cost Western European coal and around 15 per cent of total production costs of comparatively cheaper United States deep-mined coal. Flue gas desulphurization with efficiencies of up to 90 per cent are estimated to cost about 15 to 20 per cent of production costs in Western Europe and up to 50 per cent of production costs for high-sulphur United States coal 3/.

1/ UN ECE, Environmental aspects of energy production and use, with particular reference to new technologies (SC.TECH/R.44).

2/ UN ECE, Report of the Symposium on environmental problems resulting from coal industry activities (Katowice, October 1976) (COAL/SEM.5/2, para.42).

3/ OECD, Energy production and environment, Paris 1977, p.78, 86, 96.

These impressive figures should not obscure the fact that the economically-recoverable reserves of low sulphur coal constitute a notable share of the coal resources of the ECE region and that accordingly the financial burden of environmental protection on the coal industry as a whole is much lower; for the Federal Republic of Germany environmental protection required 1.5 to 1.8 per cent of total production costs and 2.2 per cent of total investment costs of the hard coal industry of that country in 1974, shares which will increase to 3.1 to 4.3 per cent and 1.6 to 3.2 per cent during 1975-1979 1/. Thus, it may be concluded that generally environmental considerations are not a limiting factor to coal expansion, neither physically nor technically nor cost-wise.

(f) Industry Structure, Lead Times and Other Factors

In their reports governments have made no allusions to possible constraints resulting from the organization of the coal mining industry in their respective country. This may be interpreted as meaning that generally no major problems are expected to result from institutional factors. However, the expansion of the industry is not insensitive to ownership patterns and organizational structures: large new mines which could draw maximum benefit from economies of scale, are not likely to be opened if reserves are owned and exploited by a great number of companies, or if coal mining companies are integrated into other industries such as electric utilities or steel works.

A major limiting factor is the lead time required to increase capacity. The short-term flexibility of production under normal economic and operational conditions is limited to a few percentage points. The opening of new mines requires at best four to seven years, and the planning and building of coal-fired power stations five years, but longer lead times must be anticipated for high-capacity installations, openings in remote or densely populated areas or if difficult geological and climatic conditions prevail.

There are further determinants of future coal production which need at least to be mentioned: land leasing policies; the ability of manufacturers to supply the required underground and surface mining equipment, preparation and conversion technologies and transport equipment; an appropriate expansion of rail and water transportation fleets, including ocean transport; the standardization of equipment; the facilitation of international trade, particularly harmonization of regulations applicable to equipment use. None of these factors has been singled out as limiting the expansion of coal production, but it is also evident that an active intergovernmental co-operation with regard to the last-mentioned areas could have only a beneficial effect on expansion programmes.

C. ANTICIPATED CHANGES OF DEMAND/SUPPLY TRENDS AND PATTERNS

In the preceding chapter, the major factors determining coal demand and supply in the ECE in the medium-term have been assessed. What will be their combined effect on coal consumption, production and trade?

Table V.7 attempts at an overall quantitative reply to this question. The following section (a) analyses this table and other information from the point of view of coal demand growth, whereas section (b) analyses coal supply growth.

1/ G. Dach, G. Schmidt, Monetäre Aufwendungen für Umweltschutz in der Industrie, unter besonderer Berücksichtigung des Steinkohlenbergbaus, Glückauf, No.112 (1976) p.96.

(a) Coal Demand Growth

Overall growth rate of coal demand in the ECE as a whole. According to government plans and programmes, coal demand will rise from 1738 mill. tce in 1973 to 2370 to 2410 mill. tce in 1985 in the ECE region as a whole, i.e. 38 per cent or 630-670 mill. tce more than in 1973. The average annual growth rate of coal demand during the twelve-year period 1973 to 1985, is 2.7 per cent, compared with 0.9 per cent during the preceding twelve-year period 1961-1973. Thus, coal demand growth is expected to increase by three times. The expected long-term growth rate of 2.7 per cent has never been achieved in the ECE region since 1950; the maximum hitherto achieved was 1.5 per cent (during 1954-1966).

Regional growth rate of coal demand. All coal-consuming regions of the ECE area will increase coal demand and will accelerate demand growth.

TABLE V.6 Average Annual Growth Rate of Coal Demand

Region	1961-73	1973-85	Increase mill. tce
Western Europe	-2%	+0.9%	+40-45
Eastern Europe including USSR	+1.9%	+2.4%	+255-290
North America	+2.9%	+4.2%	+336
ECE region	+0.9%	+2.7%	+630-670

Western Europe, which presently shows a continued high reliance on imports of oil will reverse the decline in coal consumption. A substantial acceleration is anticipated in the United States which under business-as-usual conditions would rely on foreign oil to an extent of 23 per cent of its total energy demand in 1985. An acceleration of the same order of magnitude is planned in the centrally-planned economies of the ECE region which reflects the expectation that the present comfortable position of this group of countries with regard to hydrocarbons might fade away as a result of the continued high rate of economic growth and growing costs and difficulties of tapping new reserves.

Projected coal demand growth in the world. How does the ECE coal demand growth rate of 2.7 per cent compare with the projected coal demand elsewhere in the world? Figures are obviously less firmly established, but judging from production forecasts and world trade patterns, it seems that also in the non-ECE countries the coal demand growth rate will go up (from 3.4 per cent during 1961-1973 to probably between 4.8 and 5.3 per cent during 1973-1985). However, the acceleration is smaller: + 50 per cent compared with + 200 per cent in the ECE region. What are the reasons for this reaction?

In a report submitted to the fifth session of the Committee on Natural Resources 1/, the Secretary-General of the United Nations points to the fact that the choice of the coal option is simply not available to many developing countries having a possible coal potential, in the absence of any significant efforts in coal exploration and development. In countries or regions with no known oil or coal or hydro resources, nuclear power is gaining ground. The petroleum-surplus countries are expanding their oil-fired power stations. A number of developing countries with accessible and comparatively inexpensive hydro-power are now giving priority to the development of hydro-power stations over thermal power from imported fuels or nuclear power, which is comparatively expensive for small plants.

However, there are now sufficient indications that the historical neglect of coal is likely to be reversed. Increased coal production has substantial employment potential. Coal is likely to play an increasing role in meeting the growing electricity requirements; the exploitation of indigenous coking coals would strengthen the establishment of the major steel industry and the validation of the iron ore output. Further outlets for coal appear in transportation (in reversing to steam locomotives; through electrification), cement and brick manufacture and ammonia manufacture. A pre-requisite for realization of these opportunities seems to be international co-operation in technical and financial terms.

For the world as a whole: total coal demand is expected to grow at a rate of 3.4 to 3.7 per cent (compared with 1.7 per cent in 1961-1973). In 1985, total world coal demand will be between 3,705 and 3.815 mill. tce i.e. about 1,300 mill. tce more than in 1973. Sixty-four per cent will be generated within the ECE region.

Change of sectoral supply patterns. Changes in the pattern of supply are in line with past experience: a relatively and absolutely considerable increase in the coal supply will take place in the power station sector; the amount required by the iron and steel and coking industries will continue to grow in absolute terms, but its share in the total coal supply remains practically unchanged; small absolute but important relative losses will affect the remainder of the market mainly accounted for by households, small consumers and in the United States by coal gasification and liquefaction plants. The trend towards concentrating the coal supply in two consuming sectors continues: 84 per cent of all supplies in 1985 will be earmarked for use by the electric power and iron and steel industries, compared with 77 per cent in 1973 and 57 per cent in 1967 (Table V.8).

1/ United Nations, Status and prospects of coal production and utilization in developed and developing countries; E/C.7/67, para.95-109.

TABLE V.7 Solid Fuel Balances in the ECE Region 1973 and 1985

	WESTERN EUROPE[b]					EASTERN EUROPE[c]					NORTH AMERICA					ECE TOTAL				
	mill tce		pattern		annual growth 1973-1985 %	mill tce		pattern		annual growth 1973-1985 %	mill tce		pattern		annual growth 1973-1985 %	mill tce		pattern		annual growth 1973-1985 %
	1973	1985	1973	1985		1973	1985*	1973	1985		1973	1985	1973	1985		1973	1985	1973	1985	
I. BALANCES[a]																				
1. Production	346	350	87	79	±0	850	1 115-1 150	104	104	+2.4	552	940	105	109	+4.5	1 748	2405-2440	101	101	+2.8
+2. Imports	79	110-120			+3.4	43	48-54			+1.4	16	15*-20*			+0.7	138	178-194			+2.5
-3. Exports	34	25	13	21	-2.5	77	93-98	-4	-4	+1.8	61	95*-100*	-5	-9	+4.0	172	213-223	-1	-1	+2.0
-4. Additions to stocks (+=withdrawals)	8	...			...	-1	...			...	17	...			...	24	...			...
=5. Total inland deliveries of which to	400	440-445	100	100	+0.9	815	1 070-1 105	100	100	+2.4	524	860*	100	100	+4.2	1 738	2370-2410	100	100	+2.7
(a) power stations	195	250	48	56	+2.1	450*	665-690	55	62	+3.5	369	655	70	76	+4.9	1 014	1570-1595	58	66	+3.8
(b) iron and steel works coke ovens	110	115	28	26	+0.4	115*	180-190	14	17	+4.0	98	120	18	14	+1.7	323	415-425	19	18	+2.2
(c) others	95	75-80	24	18	-1.7	250*	225	31	21	-1.0	57	85	12	10	+3.4	402	385-390	23	16	-0.2
II. INDICATORS																				
1. Share of coal in energy supplies			26	20				46	36				19	22				28	26	
2. Share of hard coal production			81	78				69	74				99	93				81	81	
3. Share of underground prod.			82	69				62	59				51	38				63	52	
4. Share of regional coal prod. in total ECE coal prod.			20	15				49	47				32	39				100	100	

a/ Hard coal, brown coal, lignite converted into tons of coal equivalent according to UN Statistical Series J "World Energy Supplies"
b/ Incl. Turkey.
c/ Incl. USSR.
*=Secretariat estimates.
...= Not available
for further explanations, see tables II.1, III.1 and IV.2.

TABLE V.8 Coal Demand by Consuming Sectors, 1973 and 1985

Region	Pattern, in % of total demand			Change 1973-1985, mill. tce			
	power stations	iron/ steel	others	power stations	iron/ steel	others	total
Western Europe							
1973	48	28	24	+55	+5	-15-	+40-
1985	56	26	18			-20	45
Eastern Europe incl. USSR							
1973	55	14	31	+215*-	+65*-		+255*-
1985	62*	17*[a]	21*[a]	+240*	+75*	-25*	+290*
North America							
1973	70	18	12	+286	+22	+28	+336
1985	76	14	10				
ECE 1973	58	19	23	+556-	+92-	-12-	+630-
1985	66*	18*	16*	581	+102	-17	+670

a/ Taking into account that steel demand in the centrally-planned economies will grow by 4.2 per cent during 1976-1980, a growth rate of 4 per cent for related demand for coking coal and steam coal has been assumed for 1973-1985; the estimated decrease, at a rate of 1 per cent of direct coal supplies to "others" is less well founded and reflects assumed losses in the railway sector (to the benefit of electric or Diesel traction) and in households (to the benefit of district heating).

The impact of an increased coal supply on the raw material pattern of power stations will be discussed under D (a) below.

(b) Coal Supply Growth

The fact that this section on the supply side of the new coal policy follows the section on the demand side, should not be interpreted as meaning that the coal supply is simply a function of coal demand. Certainly "supply" follows "demand" if the challenge on the supply side is not extraordinary in terms of production growth, resource depletion, manpower needs, lead times. But such a statement would not seem representative for the situation everywhere in the ECE region and for the specific situation which prevailed after 1973/74. Too often coal demand proved supply-constrained, certainly in the short term and if a major rise of production was aimed at. Easing constraints on the supply side therefore became an equally important task for ECE governments as easing constraints on the demand side. With these considerations in mind, this section will analyse the supply side of the new coal policy not just as a mere function of rising demand, but also as a constraint to demand.

Supply strategy. It appears from government plans and projections that the total energy demand in the ECE region will rise between 1973 and 1985 from 6,064 mill. tce to 9,300 mill. tce, i.e. by approximately 3,230 mill. tce. The contribution of coal towards meeting the incremental energy demand in the various ECE regions will be 6 per cent in Western Europe, 22 per cent in Eastern Europe and the USSR, 39 per cent in North America and 20 per cent in the ECE region as a whole.

Several strategies have been chosen by ECE governments to procure the additional coal supply estimated at 630 - 670 mill. tce, i.e. by:

- increased production;
- increased imports;
- reduced exports.

Table V.10 shows that the coal supply strategies chosen by the three major ECE regions are quite different. One might have expected that an increase in production would be the most important avenue for meeting increased demand. This is true for the ECE region as a whole where the increasing coal production (657-692 mill. tce) will be bigger than the increase in coal demand (630-670 mill. tce). This will be true also for North America and Eastern Europe where production growth will cover all the internal coal demand increase as well as new foreign requirements. But for Western Europe, the most important parameter for procuring the required new coal supplies of about 40-45 mill. tce, will not be production, but imports; for this region, the rise in imports (+ 36-41 mill. tce) will be such that the contribution of indigenous production to the total coal supply will fall from 87 per cent in 1973 to 79 per cent in 1985.

These different supply strategies both as concerns the choice between coal on the one hand and other forms of energy on the other, and as regards the relative weight given to indigenous coal production and coal imports, are important indicators of the medium-term competitiveness of coal in Western Europe, Eastern Europe and North America, the first ranging at the lower end, the latter at the higher end.

Overall growth rate of coal production in the ECE as a whole. In 1985 production of hard coal and brown coal in the ECE region will be higher by about 38 per cent than in 1973 and reach 2,405-2,440 mill. tce. The net increase in coal production is about 660-690 mill. tce. The annual growth rate of coal production during the twelve-year period 1973 to 1985 is 2.8 per cent and compares with 0.8 per cent in the preceding twelve years. Thus, coal production growth is expected to accelerate three to four times. The anticipated medium-term growth rate of 2.8 per cent has never been achieved in the ECE region after the Second World War.

Production growth by type of coal. In the ECE region as a whole, production of hard coal and brown coal(on a tce-basis) will grow at practically the same rate of 38 per cent. This balanced growth pattern is due to the extraordinary development of hard coal production in the eastern parts of the USSR and to the steep increase in brown coal/lignite mining in Southern Europe, the Balkans and North America. Production of hard coal will grow by about 550 mill. t and production of brown coal/lignite by 125 mill. tce or 260 mill. t in the ECE region as a whole. The share of brown coal/lignite will remain at 19 per cent. Eastern Europe will remain the most important producer of

TABLE V.9 Supply Strategy to Meet the Increased Energy and Coal Demand Between 1973 and 1985

(mill. tce)

Supply strategy	Western Europe	Eastern Europe incl. USSR	North America	ECE
I. Energy supply strategy				
(a) total energy demand 1985	2 275	3 014	4 013	9 302
(b) increased energy demand 1973-1985	708	1 236	1 294	3 238
(c) increased coal demand 1973-1985	40-45	255-290	336	630 - 690
(d) (c) in per cent of (b)	6	22	39	20
II. Coal supply strategy				
(a) total coal demand 1985	440-445	1 070-1 105	860	2 370-2 410
(b) increased coal demand 1973-1985	40-45	255-290	336	630 - 670
(c) (b) covered by				
(i) increased production	4	265-300	388	657 - 692
(ii) increased imports	36-41	5-11	-1- +4	40-56
(iii) reduced exports (- = increased exports)	9	-16- -21	-34- -39	-41- -51
(iv) correction for 1973 withdrawals from stocks	-8	+1	-17	-24

brown coal/lignite (1973: 79 per cent of ECE total; 1985: 65 per cent), but brown coal/lignite mining is expected to undergo a fast absolute and relative growth in North America (1973: 2 per cent of ECE total; 1985: 17 per cent). For further details see Tables V.10 and V.7 (line II 2).

Production growth by mining technique. The share of open-cast mined coal in total production will grow from 37 per cent in 1973 to 48 per cent in 1985, i.e. almost half of the 1985 production will be by surface mining. Out of the new coal production of about 660-690 mill. tce, 75 per cent will be by open-cast mining - an extraordinary challenge on equipment manufactures. As can be seen from Table V.7 (line II 3), all ECE regions expand surface mining. In Southern Europe, almost all the increase in coal production will come from such mines.

Regional production growth. All ECE regions anticipate a growth of coal production. In all regions the growth rate lies above that suggested before the oil crisis. But the rate anticipated for the various ECE regions is quite different:

+ 0.01 per cent for Western Europe for 1973-1985 compared with -3.1 per cent for 1961-1973;

+ 2.4 per cent for Eastern Europe including the USSR, compared with 2 per cent;

+ 4.5 per cent for North America, compared with 3 per cent.

TABLE V.10 Production of Hard Coal and Brown Coal/Lignite, 1973 and 1985

Region	1973					1985				
	Hard coal mill tce	Brown coal mill tce	Brown coal mill t	Total mill tce	Total mill t	Hard coal mill tce	Brown coal mill tce	Brown coal mill t	Total mill tce	Total mill t
Western Europe	284	62	180	346	460	255	95	260	350	515
Eastern Europe incl. USSR	587	263	647	850	1 235	842	290	720	1 115–1 150	1 620–1 670
North America	546	6	17	552	565	870	70	185	940	1 055
ECE	1 417	330	844	1 748	2 260	1 967	455	1 105	2 405–2 440	3 190–3 240

Thus, the new coal production policy does not appear so much as an extention of the earlier ones, but as a correction, the correction being the greater the greater the loss of market shares of coal in the total energy balance during 1961-1975.
As a result of the difference in the growth rate, the relative importance of the various ECE regions as coal producers will change: Eastern Europe, including the USSR remains the most important region accounting for 47 per cent of total ECE coal production in 1985 (1973: 49 per cent). North America will increase its weight substantially from 32 per cent to 39 per cent, while Western Europe will account for only 15 per cent of the 1985 coal production in the ECE area compared with 20 per cent in 1973. The relative dislocation of coal production to more distant coal fields has already been analysed in section 3 (b) above.

Projected coal production growth in the world. How does ECE coal production growth by almost 700 mill. tce compare with prospects for coal elsewhere in the world by 1985? Table V.11 shows that the coal producing non-ECE countries of the world announced increases of production by between 560 and 640 mill. tce, the tonnage reached in 1985 being between 1,300 and 1,375 mill. tce. The annual growth rate which had been 3.4 per cent during 1961-1973 will increase to 4.8 to 5.3 per cent during 1973-1985. This acceleration is smaller than the one anticipated for the ECE region, for reasons explained in section C (a) above.

TABLE V.11 World Coal Production 1973 and 1985
(in mill. tce)

Country	1973	1985	Source for 1985 figures
Australia/ New Zealand	65.4	104	OECD, World Energy Outlook, p.93.
Japan	22.4	18.1-20	ibid., p.92; World Coal, May 1976
India	78.3	210	Sixth Five Year Plan target
China	428	670-730	1961-1973 growth rate
Indonesia	0.2	30	Revue de l'énergie, Feb.1977,p.68
Latin America	4.1	20	ibid.
Above countries	598.4	1,052-1,114	Growth rate: 4.8-5.3
Other non-ECE countries	141	248*-261*	Growth rate assumed as above
Total non-ECE	739	1,300-1,375*	
ECE	1,748	2405-2440	Growth rate: 2.8
World	2,487	3705*-3,815*	Growth rate 3.4-3.7

Taking ECE and non-ECE countries together, coal production trends for the world as a whole would seem to be as follows: in 1985 total production of hard coal and brown coal on a tce basis in the world will be between 3,705 and 3,815 mill. tce. This is an increase of between 1,218 and 1,328 mill. tce. The growth rate will be 3.4 to 3.7 per cent compared with 1.7 per cent in 1961-1973. The relative importance of the ECE region as a coal producer will continue to fall from 90 per cent in 1950 and 70 per cent in 1973 to 64 per cent in 1985.

Overall growth of exports from, imports to and intra-regional trade within ECE. In 1973 ECE countries imported 138 mill. t and exported 172 mill. t of coal and coke, i.e. the ECE region was a net exporter of coal to an extent of 34 mill. t. This net "surplus" was the result of the net export position of North America (both Canada and the USA) and of Eastern Europe (Poland, USSR, Czechoslovakia, GDR), outweighing net import requirements from Western Europe. Intra-ECE coal trade was at 134 mill tce.

TABLE V.12 Coal Trade 1973 (mill. t)

Imports by	Exports from					
	Western Europe	Eastern Europe incl. USSR	North America	ECE	Rest of the world	World
Western Europe	30.9	29.7	13.3	73.9	4	77.9
Eastern Europe incl. USSR	0.8	41.4	0.3	42.6	0.2	42.8
North America	0.9	-	16.2	17.1	-	17.1
ECE	32.6	71.1	29.8	133.6	4.2	137.8
Rest of the World	1.2	6.6	31.3	38.8	27.9	66.7
World	33.8	77.7	61.1	172.4	32.1	204.5

Source: United Nations, Statistical Series J, World Energy Supplies 1950-1974, Table 4

By 1985, coal imports by ECE countries and coal exports from ECE countries will grow by between 40 and 56 mill. tce. The region will remain a net exporter of coal of between 20 and 45 mill. t. North America will increase net exports by between 30 and 40 mill.t, and Eastern Europe by between 5 and 15 mill. t, whilst Western Europe will increase net imports by 45 to 50 mill. t. Intra-ECE coal trade will certainly grow substantially, but figures supplied do not allow to quantify the increase.

World coal trade by volume and type of coal. Total world coal trade was around 205 mill. t of coal and coke in 1973, of which only 28 mill. t were extra-ECE trade, i.e. was neither shipped from nor to ECE countries. This trade concerned mainly shipments from Australia to the Far East, and from South Africa.

OECD 1/ estimates that the increase of net exports by countries other than ECE countries will be 33.5 mill. t till 1985, with Japan requiring the bulk. The increase of total world coal trade can therefore be estimated to be between 70 and 82 mill. t, or 37 per cent; the total tonnage shipped in 1985 will be between 275 and 285 mill. t, of which 240 to 250 mill. t touching ECE.

Slightly higher figures of 285-305 mill. t have been suggested by Anderheggen 2/ on the basis of an assumed 2.5 per cent annual growth of coking coal demand in the world, and of a 5 per cent growth of steam coal demand by OECD countries. Trade flows are anticipated to be as follows:

TABLE V.13 World Coal Trade in 1985, Total and by Type of Coal

(in mill. t)

Country	Import requirements			Export availabilities		
	coking coal	steam coal	total	coking coal	steam coal	total
Japan	85-100	15	100-115	-	-	-
European Economic Community (incl. intra-EEC trade)	35	35	70	10	10	20
Other western Europe	15	5	20	-	-	-
CMEA	10	35	45	30	60	90
Canada	10	10	20	20-25	5	25-30
USA	-	-	-	40-90	10-40	50-130
Australia	-	-	-	50	0-20	50-70
South Africa	-	-	-	5	20	25
Others	30-35	-	30-35	5	10	15
Total	185-205	100	285-305	160-215	115-165	275-380

No trade in synthetic gases and liquids from coal is expected to develop before 1985.

D. POLICY APPRAISAL

After the above analysis of the anticipated coal demand/supply developments, it is now possible to evaluate the extent to which these developments contribute to the achievement of the policy objectives which ECE governments had in mind when re-formulating their coal policies after 1973-74. It is recalled (see section A) that these policies aimed mainly at a greater security of supply, a certain control of the price developments on the energy markets and at "gaining time" to reconsider existing and implement new general energy policies.

1/ OECD, World Energy Outlook, Paris 1977.

2/ European Coal Information Agency, News and Comments, No.11/1976.

(a) Diversification

Primary energy supply. Assuming that "security of supplies" can be assimilated to "diversification of supplies", calculations have been made in Table V.14 to evaluate the role of coal in the total energy economy of ECE and of its sub-regions under various assumptions.

In 1973, the share of coal in total primary energy supplies of the ECE region was 28 per cent. Without a change in coal policies, the share of coal would have fallen to 18 per cent in 1985, in Western Europe even to 10 per cent (alternative I).

TABLE V.14 Diversification of Energy Supply: The Share of Coal in 1973 and 1985

Region	1973 actual	1985		
		Alternative I: extrapolation of 1961-1973 trend	Alternative II: as I but with coal supply growing as en-visaged by governments	Alternative III: as II but with growth rate of total energy demand slowing down a/
Western Europe	26	10	16	20
Eastern Europe incl. USSR	46	31	33	36
North America	19	16	19	22
ECE	28	18	22	26

a/ As estimated in Parts II to IV of this study, namely for Western Europe a slow-down from 5.1 per cent (1961-1973) to 3.2 per cent (1973-1985), for Eastern Europe from 5.4 per cent to 4.5 per cent, and for North America from 4.5 per cent to 3.3 per cent.

However, governments have changed their policies as a result of which coal will play a greater role than anticipated before the "oil crisis". The extent of coal's additional contribution depends not only on the tonnage of coal produced and supplied, but also on the future rate of growth of energy demand: if energy demand continues to grow at the historical 1961-1973 rate, which was 4.9 per cent for the ECE region as a whole, coal's share in the total ECE balance in 1985 would be 22 per cent (alternative II). If energy demand will grow at a reduced rate resulting from a slowdown of economic growth and greater energy economy and efficiency, the share of coal in 1985 would be 26 per cent. This alternative III is considered more likely than others. Coal's projected share of 26 per cent in 1985 compares with 28 per cent in 1973. Only in North America would coal's future share be bigger than the present one.

Thus, the new coal demand policies as they took shape after 1973/74, substantially contribute to the diversification of energy supplies in all ECE regions if compared with pre-crisis policies. However, they are not sufficiently vigorous to maintain the level of diversification which actually existed in

1973 in Western and Eastern Europe. Maintaining the 1973 level of 28 per cent would either require a further acceleration of coal demand or a further slowdown of energy demand or a combination of both. Unless such a further review of coal policies takes place, particularly oil supplies would have to be increased, not only in absolute terms, but would also have to take over market shares from coal in Western and Eastern Europe.

Developments elsewhere in the world are similar to those in the ECE region: in spite of a notable acceleration of coal demand (from 4 per cent annually in 1961-73 to 5 per cent in 1973/85) the share of coal is likely to fall from 42 per cent in 1973 to 40 per cent in 1985 1/. In the world as a whole, coal will cover in 1985 approximately 30 per cent of total energy requirements compared with 32 per cent in 1973.

Secondary energy supply. In security-of-supply policies, power generation and distribution plays a particularly important role as a buffer between primary energy supplies and final energy consumption. One might even measure the absorptive capacity of an energy economy against disruptions of primary energy supplies in terms of electricity use, relative to total final energy consumption: the absorptive capacity of the energy economy is the bigger, the greater the share of electricity in total consumption, the greater the stocks held by the electric utilities and the greater their capability of switching from one fuel to another.

The new coal policies increase security of supplies in that they increase the available raw material basis for electricity generation until 1985 (see Table V.8)

- in Western Europe by 55 mill. tce;
- in Eastern Europe by 215-240 mill. tce;
- in North America by 286 mill. tce;
- in the ECE region by 556-581 mill. tce.

From the point of view of the coal industry, these increases constitute an impressive effort: indeed, the total additional tonnage made available to the electric power industry of the ECE region as a whole by 1985, corresponds to more than the total coal production of North America in 1973. However, this increase is not sufficient to maintain the share of coal as an input for power generation: the growth rates of coal supplies to power stations however big and impressive they may be from the point of view of the coal industry, lie below the growth rates of electricity demand in Eastern Europe, Western Europe and the ECE region as a whole; only in North America might the share of coal, as an input for power generation, increase and hence contribute to the diversification of the supply base.

1/ At the estimated rate of growth of total energy demand of 6 per cent for the countries outside the ECE region, of 3.6 per cent for the ECE region and of 3.9 to 4.2 per cent for the world as a whole.

TABLE V.15 Growth Rate of Coal Supply to Power Stations, and of Electricity Generation, 1973-1985

Region	Growth rate of coal supply to power stations	Growth rate of total electricity generation	Source
Western Europe	2.1	5.6	OECD, op.cit., table A4
Eastern Europe, incl. the USSR	3.5	6*-7*	
North America	4.9	4.2	OECD, op.cit., tables A2 and 3
ECE	3.8	5.2*	

Thus the concentration of coal supplies to power stations does not provide a potential for substitution away from oil, nuclear power, natural gas or other energy carriers in Europe, and provides only a marginal improvement in the United States (from 44.5 per cent in 1974 to 45.4 in 1985) 1/. On the contrary, these forms of energy taken together will have to cover a greater share in 1985 than in 1973. However, in absolute and relative terms, this contribution of the "other" forms of energy, will lie substantially below what had been expected before the "oil crisis". It is with respect to these pre-crisis plans and policies that the new 1985 supply pattern of power stations in the ECE region can be qualified as less dependent on (imported) oil and nuclear power.

(b) Pricing

One of the motivations underlying the re-appraisal of the role of coal subsequent to the oil crisis of 1973/74, had been the desire to increase the capability of the energy economies to resist against sudden and extraordinary increases of the energy price level and hence to protect balance-of-payment equilibria, employment and economic growth. To what extent will the increase of coal supplies as described above contribute towards smoothening or preventing exposure of ECE energy economies to such price increases?

For obvious reasons the answer to this question will remain general and speculative. Indeed forecasts of the price movements of coal relative to competing forms of energy would seem very difficult:

- prices depend on a multitude of factors amongst which external factors such as government interventions, business fluctuations and installed equipment may play an overriding role in the short and medium term;

1/ Federal Energy Administration, National Energy Outlook, Washington 1976 p. 241.

- price movements are not uniform either as regards "coal" nor as regards its competitors. They differ according to the type of coal (hard coal, brown coal), the markets (power stations, iron and steel industry, households) and the institutional, organizational structure of the competing industries, in particular their capability of conducting a transnational and a multi-product price strategy.

Thus, it appears out of order to speculate about future price relationships. However, attempts seem justified and have indeed been made to evaluate the long-term relationships of production costs of coal relative to competing forms of energy, particularly nuclear power, oil and natural gas. The assumption underlying such considerations is that in the long run the relative physical scarcities of the various forms of energy become the decisive factor for their extraction costs and hence total costs, and that these relationships could not be ignored continuously during price formation. Taking into account that the economically recoverable reserves are much bigger in the case of coal than in the case of oil, natural gas and uranium, this school of thinking concludes that the production costs of coal will fall relative to those of oil, natural gas and uranium. It would depend on the importance of production costs in total costs and of the relation between total costs and market prices when, and to what extent, coal becomes cheaper than its competitors.

An attempt to quantify this argument has been undertaken in a United Nations report entitled "The future of the world economy" prepared under the direction of Professor W. Leontief (ST/ESA/44). This study, the particularities of which have to be well understood to correctly assess its conclusions 1/ relates the cost-based prices of coal, oil and natural gas to the general price level

---

1/ The study aims at identifying the consequences on the world economy of a reduction of the difference between per capita income in developing and developed countries by half during the period 1970-2000. Therefore, the figures are not forecasts of what is likely to happen, but simulations of what would happen if the objective of cutting the income gap by half would be achieved. The calculations are based on production costs (i.e. exclusive of upgrading/ conversion, transportation) and were derived from one region only (North America); it was assumed that this region produces and absorbs such a substantive part of the world total marketable output that figures would be representative for the world as a whole. The study admits that additional computations would have to be made for other regions. The changes of the relative production costs of the various forms of energy and of the other commodities are measured against the average prices of final consumption, which are kept unchanged. The changes in relative prices reflect the expected shifts in the techniques of production, in labour productivity and in resource extraction relative to the average prices level (kept unchanged). Mineral resources (copper, bauxite, nickel, zinc, lead, iron ore, petroleum, natural gas, coal and other resources) would become more expensive to an extent of 2.67 times between 1970 and 200, agricultural products to an extent of 14 per cent, whilst the average price for manufactured goods would decline by 6.8 per cent. It can thus be concluded that in the scenario described, the costs of coal production in absolute terms would rise at a rate slightly below the general price increase, but slower than the costs of oil and natural gas.

of the economy and concludes that relative to this general price level, cost "prices" for coal will be slightly falling, oil "prices" will be increasing approximately three times, and "natural gas prices" will be increasing approximately six times between 1970 and the year 2000. It may be added that also the cost relationship between nuclear power and coal is expected to develop in favour of coal. Thus, to the extent that physical scarcities affect earlier or later production costs, and that production costs affect earlier or later prices, the acceleration of coal supplies will exercise generally a moderating influence on price increases for energy. This moderating influence will gain momentum in the 1980s, but its full impact will only be felt from the late 1980s onwards.

TABLE V.16 The Future Impact of Physical Scarcities on the Relative Prices of Coal, Oil and Natural Gas

Form of energy	Share in reserves	Change in relative prices a/		
		1980	1990	2000
Natural gas	9	1.54	5.45	7.56
Oil	19	1.33	2.89	3.25
Coal	72	0.95	0.87	0.86

a/ 1970=100.

Source: ST/ESA/44, table 62.

(c) Substitution

The expansion of coal supplies particularly to power stations provides governments with an opportunity to consider their medium-term energy supply policies under less strain than otherwise. These policies require reconsideration because one major element - nuclear power - had become increasingly surrounded by uncertainties. Between 1974 and 1977, the nuclear capacity expansion programmes for the year 1985 had been substantially slowed down. This slowdown was motivated by lower electricity demand growth rates, by escalating construction costs, longer lead times, a reduced margin versus fossil fuel power generation, by growing public resistance against nuclear power in certain countries and by the slowdown of economic activity in certain countries.

The preceding section (a) had shown that in spite of increased supplies to power stations, coal will not be able to maintain its relative importance as a raw material for power generation in Europe, or to increase its market share substantially in the United States. These consequences of the new coal policies are of utmost importance for the determination of the medium-term role of (imported) oil, nuclear power and natural gas. The new coal policies provide a breathing time for the reconsideration of the supply mix, the development of new technologies, the implementation of energy conservation policies, but not for a fundamental change. The quantitative importance of the breathing time is difficult to assess, but if coal supplies would have remained at the 1973 level, other fuels (probably imported oil in the short run and nuclear power in the medium run) would have to supply the total incremental requirements of electric utilities of 380 mill. tce 1/ instead of about 335 mill. tce in Western Europe, and of 540 1/ mill. tce instead of about 255 mill. tce in North America.

E. CONCLUSION AND OUTLOOK BEYOND 1985

What is the message which emerges from the analysis of the new coal policies and programmes established 96 ECE governments after 1973/74 and illustrated in figures 1-12 below?

The main thrust is that there is plenty of coal of the required qualities (high calorie, coking, low sulphur) or suitable for cheap (opencast) mining. Without undue prejudice for future generations the economically recoverable coal reserves in ECE and its major sub-regions could sustain a rate of growth of coal production and international coal trade which goes beyond the rate of growth of energy demand, and hence would reduce reliance on other forms of energy.

However, the actual developments will be different. This is less the result of a genuine incapability of the coal industry to live up to expectations, although geological conditions will worsen to the extent that old coal fields have to be abandoned, that production will have to be dislocated to "distant" coal fields and that lesser countries than before will be able to supply the full range of coal qualities. The main reasons for the divergence between potential and actual developments would seem to be temporary and local in nature, and could be overcome in the longer term and through trade, transportation and improved international co-operation.

What are these constraints? The competitive advantage which coal had gained in 1973/74 as a result of the fourfold increase of crude oil prices turned out to be less striking in the short-term: coal prices rose, too, in order to reduce losses/government subsidies and to improve profitability; and not everywhere was the increase of crude oil prices equally reflected in the price of heavy fuel oil with which coal competes. The medium-term flexibility of the coal industry (as of any heavy industry) was low: lead times for the adaptation or construction of production and conversion facilities and for the development and penetration of new technologies were notable. Limitations appeared as a result of the need to reconcile the new coal policies with other government policies: employment and social policies which affect the recruitment of new labour; investment policies which affect expansion and productivity; environmental policies which affect coal mining/transportation/conversion techniques and costs. These constraints in particular the limited growth potential of labour productivity, the lack of capital for coal developments for exports, the need for recultivation and desulphurization turned out to be limiting factors in the medium-term.

Accordingly, coal demand growth in the ECE region in the period 1973-1985 will remain below the potential suggested by resource availability and patterns. Coal demand growth will be at 2.7 per cent annually, an acceleration of three to four times compared with the preceding 12-year period 1961-1973. The constraints mentioned above are reflected in different growth paths of the three major ECE regions: Western Europe + 0.9 per cent, Eastern Europe and the USSR +2.4 per cent, North America +4.2 per cent. Whilst impressive from

---

1/ OECE, World Energy Outlook, Paris 1977, tables A2, A3 and A4.

the point of view of the coal industry, these growth rates do not suffice to maintain the share of coal; there is a decrease from 28 per cent in 1973 down to 26 per cent in the case of the ECE region as a whole, a decrease from 26 per cent to 20 per cent in Western Europe, and a decrease from 46 per cent to 36 per cent in Eastern Europe and the USSR. Only North America expects a (slight) increase (from 19 to 22 per cent). The same is true for the use of coal by power stations: in spite of a 56 per cent increase of coal supplies to power stations, the share of coal as an input will fall in Europe and stagnate in the United States, thereby excluding (in comparison with 1973) any substitution away from oil, natural gas and nuclear power.

Supply growth will be compatible with demand growth. The ECE region as a whole remains a net exporter of coal of about 20-45 mill. t. North America and Eastern Europe, including the USSR, will raise their net exports by approximately the same amount as the increase of net import demand by Western Europe. All ECE regions will increase production (Western Europe + 4 mill. tce, Eastern Europe and the USSR +265-300 mill. tce, North America + 388 mill. tce, ECE 657-692 mill. tce). 76 per cent of this increase will be in "distant" coalfields. The share of production from opencast will grow in all regions, in ECE from 37 per cent in 1973 to 48 per cent in 1985. Production of brown coal (lignite) on a tce-basis will grow at the same rate as production of hard coal; its share in total ECE production will remain at 18 per cent.

Coal production elsewhere in the world will grow faster than in ECE (4.8-5.3 per cent compared with 2.8 per cent), but the acceleration is slower. Between 1973 and 1985 world coal production will grow from 2,487 mill. tce to 3,705-3,185 mill. tce i.e. about 1,218-1,328 mill. tce or 3.4-3.7 per cent. The share of ECE coal production in world coal production will fall from 70 to 64 per cent. World coal trade will be around 275-285 mill. t, i.e. 70-80 mill. t or 37 per cent higher than in 1973; only 10-12 per cent of this trade will be extra-ECE. In view of the available export potentials, no tensions need occur in meeting import requirements, total and of steam and coking coal, if long-term arrangements will be made.

For the ECE region as a whole, for its sub-regions and - by the way - for the rest of the world, coal will not emerge as a deus ex machina capable of reducing the vulnerability against supply disruptions or extraordinary price increases. On the contrary, the continued loss of market shares by coal in the primary and secondary energy markets will even give added urgency to these questions. The new coal policies will, however, provide important relief: without them, coal would have been reduced to a form of energy "amongst others"; its share in the total ECE energy balance would have fallen to 18 per cent, in Western Europe even to 10 per cent. In slowing down the decline of coal's relative importance, or (North America) even in reversing the trend, important partial solutions become possible and long-term options are kept open: in the electric utility market, as a means of earning foreign currency, by moderating price increases on the energy markets.

The conclusion that the new coal policies enhance coal's role relative to pre-crisis plans but not in comparison with the 1973 situation, induces the question whether the constraints on coal expansion mentioned could not be overcome in the longer-term. Apart from pointing to the potential of coal gasification/liquefaction, governments have not addressed this issue. But Leontief has indicated that the scarcity of hydrocarbons relative to coal will inevitably, particularly after 1985, improve coal's competitiveness. In the present study reference has also been made to new technologies and management techniques which carry promise of relieving constraints on labour productivity, without requiring too big a contribution from the capital markets. It would seem to be the essential role of international co-operation to assess the potential of such new opportunities and to shorten lead times, through exchange of experience, joint R&D, trade promotion, standardization, investment sharing, joint infrastructure projects such as long-distance transmission lines, coal slurry pipelines, coal loading/unloading facilities, unit trains etc. The prospects for successful co-operation are bright: the similarity of problems and approaches is striking. Whilst coal policies had been in the past an example of variety and even divergence, they have appeared after 1973/74 as an example of similarity and convergence.

Topics which appear worthwhile of intensified intergovernmental co-operation within ECE include:

- the long-term prospects of coal beyond 1985
- international trade in coking coal, steam coal, low-sulphur coal
- coal mining techniques for the 1990s with particular reference to labour productivity and efficient resource use
- coal conversion by the end of this century - options, pre-requisites
- coal production growth and the manufacture of and trade in coal industry equipment
- transport and transmission problems linked with the development of "distant" coal fields
- transfer of technologies to countries with newly developing coal industries
- investment sharing in coal production, conversion, research and development, and trade
- coal industry structure and management beyond 1985 (ownership, size, multi-national strategies, concentration and combination of operations, flexibility in the face of demand fluctuations, etc.).

COAL 1985

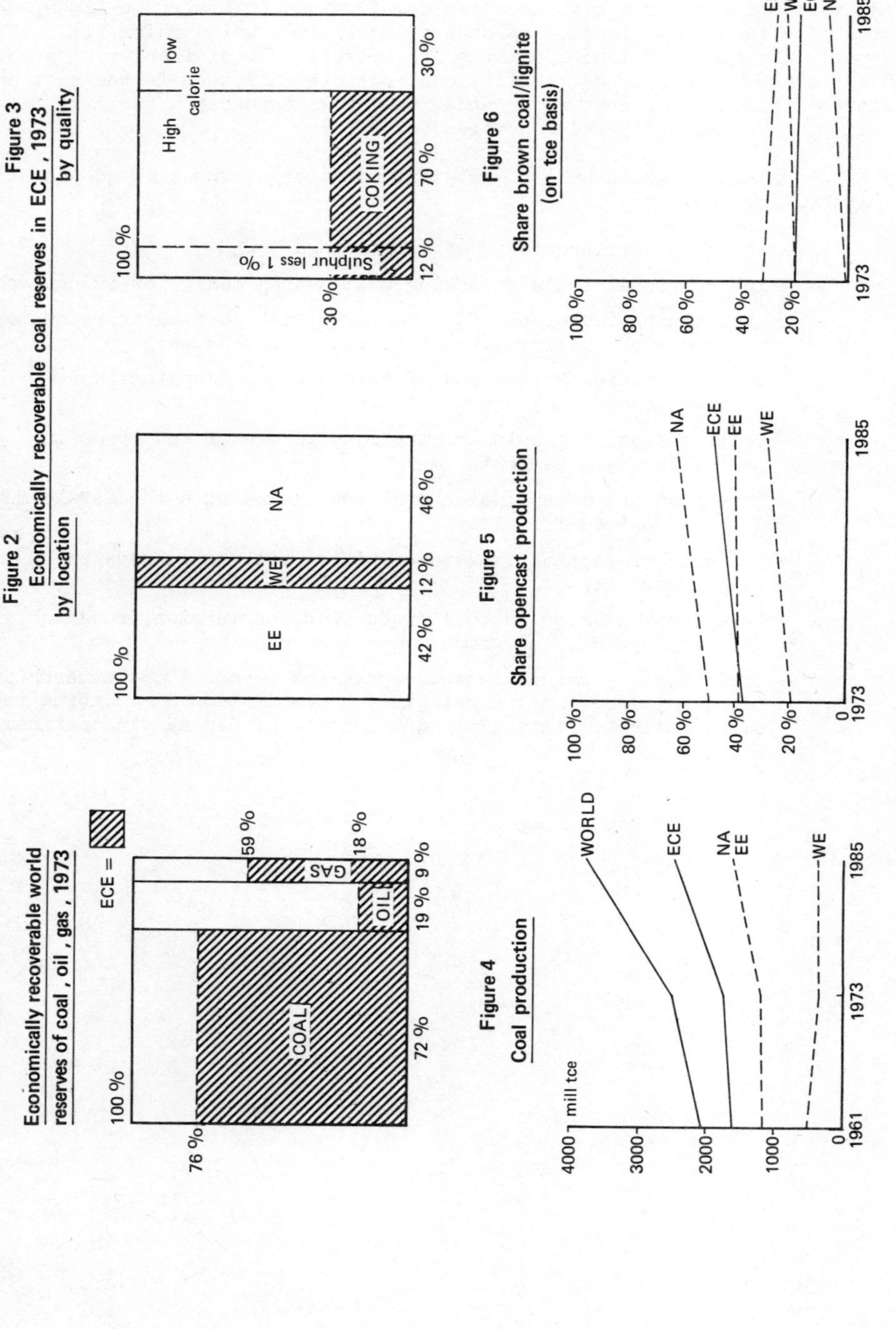

WE = Western Europe ; EE = Eastern Europe and USSR ; NA = North America

Figure 7

Production cost-based prices of coal relative to the general price level (kept unchanged), oil and gas

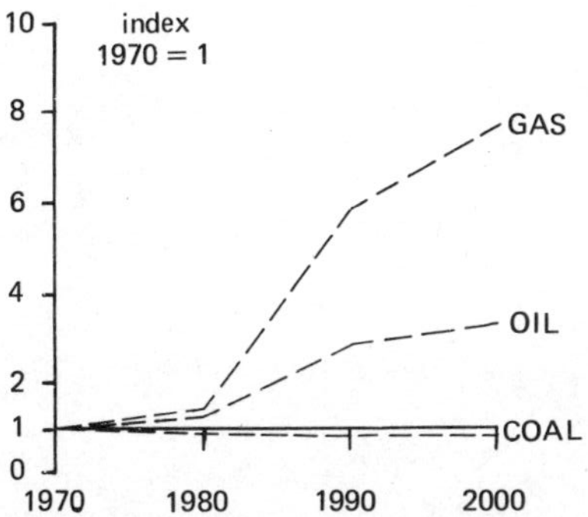

Figure 8

Net exports (+), net imports (–)

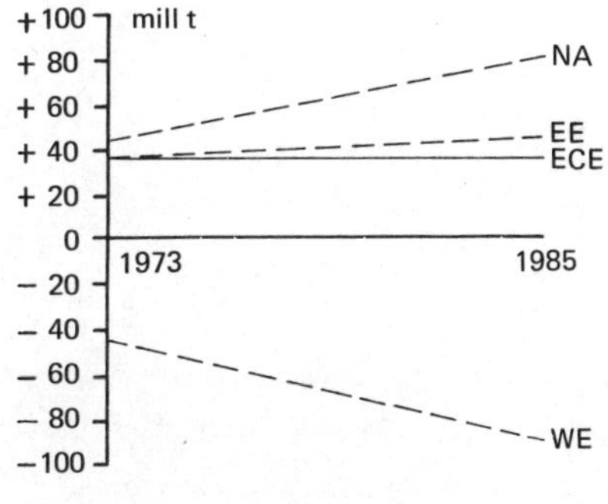

Figure 9

Coal delivery pattern, ECE

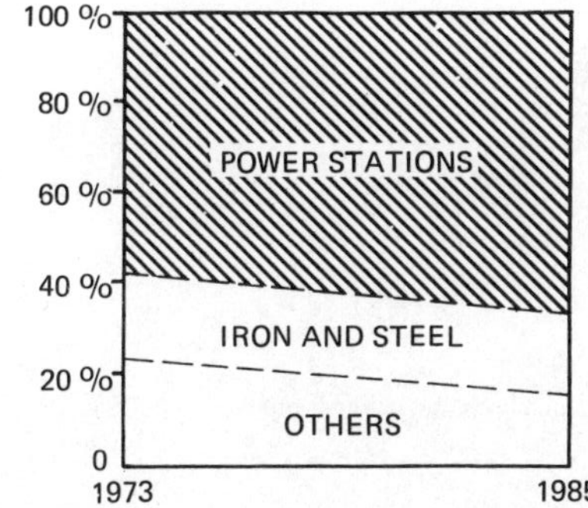

Figure 10 **Figure 11**

Share of coal in energy balances

pre-1973

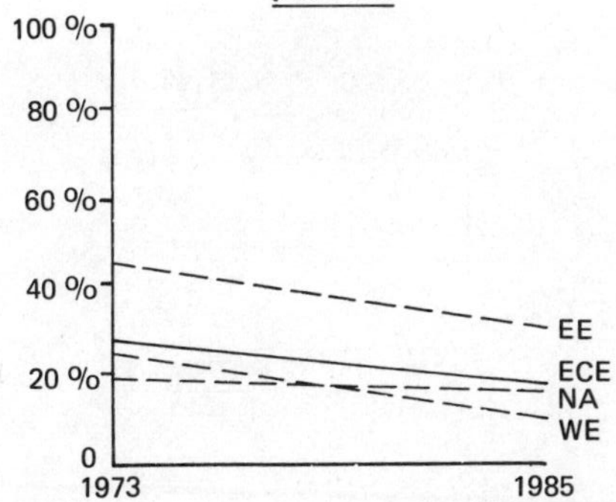

post-1973

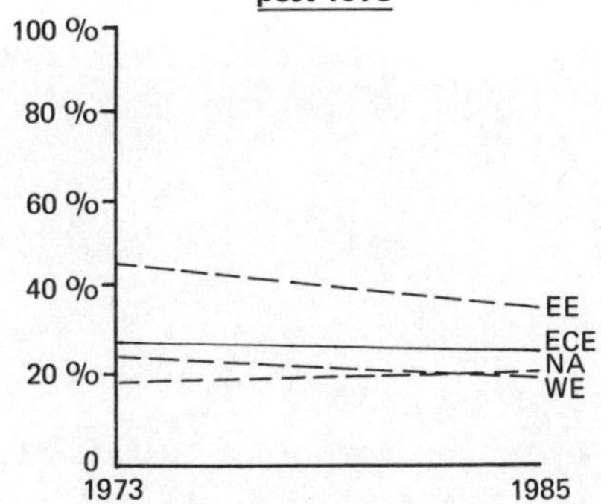

Figure 12

Growth rate of coal supplies to power stations relative to growth rate of electricity generation

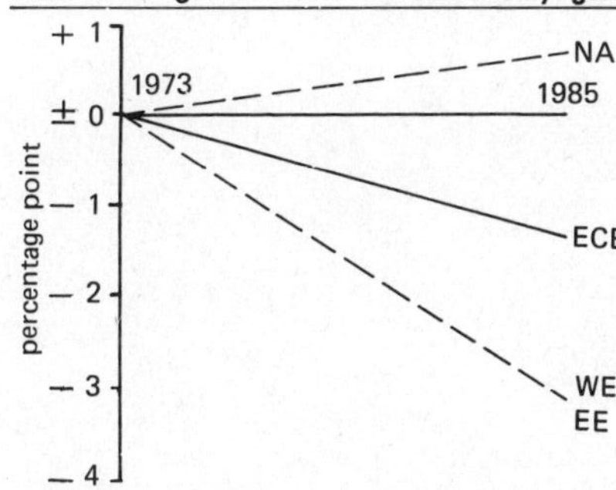